Delivering Sustainable Buildings

Delivering Sustainable Buildings
an industry insider's view

Mike Malina
Director
Energy Solutions Associates

A John Wiley & Sons, Ltd., Publication

Blackwell Publishing was acquired by John Wiley & Sons in February 2007. Blackwell's publishing program has been merged with Wiley's global Scientific, Technical and Medical business to form Wiley-Blackwell.

Registered Office
John Wiley & Sons, Ltd, The Atrium, Southern Gate, Chichester, West Sussex, PO19 8SQ, UK

Editorial Offices
9600 Garsington Road, Oxford, OX4 2DQ, UK
The Atrium, Southern Gate, Chichester, West Sussex, PO19 8SQ, UK
2121 State Avenue, Ames, Iowa 50014-8300, USA

For details of our global editorial offices, for customer services and for information about how to apply for permission to reuse the copyright material in this book please see our website at www.wiley.com/wiley-blackwell.

Library of Congress Cataloging-in-Publication Data

Malina, Mike.
Delivering sustainable buildings / Mike Malina.
      p.   cm.
   Includes bibliographical references and index.
   ISBN 978-1-4051-9417-4 (pbk. : alk. paper)   1. Sustainable building–Design and construction.   I. Title.
   TH880.M35 2013
   690.028'6–dc23

                                                                    2012027832

A catalogue record for this book is available from the British Library.

Wiley also publishes its books in a variety of electronic formats. Some content that appears in print may not be available in electronic books.

Cover images courtesy of Mike Malina
Cover design by Sandra Heath

Set in 10/12pt Palatino by SPi Publisher Services, Pondicherry, India
Printed and bound in Malaysia by Vivar Printing Sdn Bhd

1   2013

Contents

v

Foreword

It was by way of appreciating, in the 1980s, the dreadful consequences of manmade climate change that we soon came to realise that the real problems lay way beyond weather, and that solutions would require fundamental changes to the way we live on this planet. The title 'sustainability' was then coined, around the new millennium, as it seemed to encapsulate the basis for the changes needed. As building engineers, we quickly saw that the major responsibility for the problem lay with our product, with buildings producing over half of the 'climate changing' carbon emissions, using large quantities of raw materials and throwing away disproportionate amounts to waste. Put simply, we had to do more, much more, with less, far, far less.

The consequences of failure are beyond imagining, threatening the very existence of a future for mankind. We saw the new millennium as a time of 'the calm before the storm', a period having the 'luxury' of both the time and the money to plan, organise and invest cost-effectively in solutions and a new order of society. A decade on, and a clear sight of the looming storm is upon us, as greedy global financial activity has brought us to the verge of bankruptcy and we can only lament how little we've taken advantage of 'the calm', when we had the means to make progress, and how it's now so much harder to do so, with higher costs and our economies in tatters.

Yet already it's becoming clear that lessons have, even now, not been learned and solutions to economic woes are once again being sought through growth (a concept idolised too often in commercial circles), which solution is but an illusion when based, as it is, on more consumption. This is the very opposite of sustainability, where the solution comes via 'creation' through the capture of incoming energy and efficiency in its use, and its distribution for the good of everyone.

So it is that we now see money being printed, in billions, to enable more consumption and more wastage, while sustainability is being side-lined, with its 'creative potential' for real growth being sacrificed. Well, we homed-in, a decade ago, on what we had to do and now it's crucial that we let nothing deter us from pursuing those objectives with the added help from more recent discoveries whose potential could help us catch up on lost time.

Scroll forward a decade and it should be inconceivable that a building would be engineered in any way that:

- requires fossil fuel, or nuclear power, to keep it comfortable to be in (instead recovered heat should warm ventilating air and domestic hot water, insulation and airtightness eliminate fabric heat losses and minimise, with the aid of shading, unwanted gains)

- does not have adequate day-lighting and efficient, effective lighting facilities
- is not equipped with 'super smart metering' that automatically adjusts electrical demand to match the most efficient available supply
- does not employ its facade and its surrounds to capture and convert the potential of solar energy – this ranges from the simplest warming from winter sunshine, through solar thermal panels for heating water and air, and on to PV for the electricity to power indoor appliances. Many, by then, should be producing considerably more energy than they have direct need of, enabling export and power for transportation. This will also come from employing plants or algae on, or in, the facades and surrounds (100 trillion watts are captured in flower nectar each year)
- is not constructed from at least 50% recycled materials and fabricated in off-site factories as low wastage, pre-tested, modules ready for 'plug-and-play' assembly at site.

Having first met Mike at the CIBSE/ASHRAE conference in Edinburgh in 2003, I've witnessed and admired his enthusiasm and determination over the pursuit of sustainability for the engineering of buildings in particular. Now, with sustainability appreciated as an imperative for the future of mankind, the need for a host of 'Mikes' and the global dissemination of their message and knowhow, becomes paramount. His book gives us much to dwell on and also offers a great deal of useful advice on how to respond to the challenge.

Terry Wyatt,
Past President CIBSE

Preface

Most worthwhile books are the product of a long period of reflection, often spanning many years. I can trace the journey that led to this work, which seeks to make a small contribution to bridging the gap between the wider issues of sustainability and the key role of sustainable building services engineering, back to my childhood. For many, the environmental and energy performance of the built environment and many of the services crucial to this process; such as pumps, fans and ductwork systems; is less well understood, and the immediate connections are not made to wider environmental and sustainability considerations. What often gets priority is finance and cost factors, which are important and are the primary drivers for many. For me, this attempt to bring together approaches to building services engineering with sustainability – without doubt, the most pressing challenge to face present and future generations – is the culmination of an eclectic range of interests which has shaped my career and life to date.

Given the importance of the subject matter, it seems strange to me that so little has so far been written with a holistic view of both the wider environmental links and sustainable building services engineering. Much has been written about 'green issues' and there is a veritable library of handbooks and texts on building services, but to the best of my knowledge this is the first full-length work devoted solely to bringing these important subjects together. There are many environmentalists, ecologists and 'new sustainability experts' as well as, of course, many building services engineers, but there are very few who cross the divide and work with both disciplines. Therefore, this process has been largely about bridging a chasm to make new connections, and the journey that brought me to do this started long ago.

It began with an early interest in earth sciences. Anything in, or under and above, the earth was a source of fascination to me. As a child, I would seize any opportunity to dig holes, explore new sites of interest or discover more about the world around me. This interest was always eclectic. I was as intrigued by soil composition – and the invertebrates that inhabited it – as I was by the constellations above us. Microscopic organisms were as absorbing as oak trees, and pebbles as intriguing as the stars. Geology and cosmology held equal sway.

In 1969, at the age of eight, I found my first fossil. This was during an age of rapid and radical social change, but my personal epiphany at that time was all about the past. How did that shell get 'frozen' forever in that piece of ancient limestone? What kind of world had it once inhabited? I started collecting rocks and fossils, and the discovery also fostered a wider engagement with history. The prehistoric, geologically captured world of fossils held my imagination, and I also benefitted from the teaching of a great-aunt who, in those far-off days when we were largely innocent of the strictures of health and safety, would take me on trips to explore quarries. I still have many of the rocks and fossils we found. It all fired my enthusiasm for learning about the natural world. My aunt also encouraged a parallel interest in the more recent past, as revealed by the archaeological record, and I participated in a number of archaeological rescue digs from the age of 12. Was there no limit to what the earth could teach us? Accordingly, my interests at school were centred around history, geography biology and what was then craft, design and technology, and as these were the subjects that engrossed me, these were the areas where I did well. The wider world was also brought home to me as my Dad had served in the Royal Navy in WWII and had covered half the globe travelling to many exotic places. His stories and recollections inspired me to find out more about the geography and history connected to these events.

At the same time as my aunt was risking life and limb to help me extract ammonites from abandoned quarries, I was also influenced by my older brother's nascent career in electronic engineering. A good eight years older than me, he introduced me to circuit boards, switches and components. This, my first exposure to the world of technology, led me to speculatively dismantle many items to see if they could be successfully reassembled. I was an inquisitive child, and also quite a determined one – I couldn't accept that broken things couldn't be fixed again. Looking back, I think that this time was also the point where the crossover between an interest in the natural world and my parallel interest in things technological began. Rather than seeing technology as a universal solution to all challenges, I was not very old when I first realised that most of our engineered solutions are miserably clumsy compared to those refined by nature over millennia. As a fully paid-up Darwinist, I don't see the wonders of nature as the creation of a higher power, but as answers to evolutionary challenges. I have long recognised that the natural world has developed all of the most elegant solutions to the problems posed by the constraints of our physical environment. We create poor copies by comparison. As such, I think I have always recognised the need to safeguard these amazing natural achievements rather than stifle them with the by-products of our own attempts at progress. Not least, we need to do this because we can learn so much from the natural processes around us, as nature's experience is way in excess of our own. Its engineering through trial and error has produced the most amazing things, and we need to base our own future solutions on our improved understanding of these structures and processes.

I never doubted that such advances were possible. Growing up in the sixties and early seventies meant living through a time of progress and optimism, when many momentous things seemed to be achievable. Between the ages of 9 and 11, from 1969 to 1972, I was allowed to stay up to watch the Apollo programme moon landings. The buzz and excitement around the events were palpable, and had a massive effect on many young children at the time. I became an avid collector of newspaper cuttings, which at the time were assuring us all that by 2000 we would have bases on the moon, and would perhaps even have been to Mars. I remember the first time that I saw the famous, now iconic, image of the astronauts looking back at the earth. It was the first time the earth had been seen pictured from afar; now that pictures of the earth from space are commonplace, it is hard to remember what impact it had. The moon landings were arguably the crowning achievement of the age, and yet their defining image is a stark reminder of our finite place in the universe. Humanity was, we must remember, driven to conquer the moon by the relentless international competition bred by the cold war, and so it was essentially an extension of the arms race. Peace was a fragile commodity back then, but the race to the moon brought us a reminder that the world, which could it seemed have been engulfed by war at any time, was but one small globe in an infinity of space. It also emphasised just what could be done with the political and technical will to achieve the most difficult of tasks.

It seems fitting, then, that 1972 also brought the first international conference on the environment, the Stockholm Conference. Despite my tender years at the time, I feel a personal connection with that event. My mother was an avid amateur radio enthusiast, and in particular an adherent of DXing. For the uninitiated, DXing (the name comes from DX, which is telegraphic shorthand for 'distance') is the practice of tuning into distant radio stations. Listeners would send in reception reports to these far-off stations, and the convention was that they would receive what was known as a QSL card in reply, which acted as a confirmation of the broadcast and an acknowledgement from the station of an accurate reception report. In the days, when there were a limited number of radio stations, and only three terrestrial television stations, this was a way to broaden one's entertainment and knowledge options by discovering a myriad of English language broadcasts from round the world. Many of the broadcasts that my mother listened to and reported on were from the Stockholm Conference, and its reporting by many stations around the world. She duly received a certificate from Radio Sweden and numerous QSL cards acknowledging her detailed participation in this important event. I can remember these broadcasts quite well, and they were the first mention I can recall of things that now dominate the agenda, such as the balance between resources and population. (This was long before the internet, when short wave was the only way of listening to foreign radio.) Just as the photograph from the moon had suggested, it seemed that the earth really was self-contained, and the resources that were present on it were finite and therefore very important to conserve and use wisely.

This interest in environmental issues was reawakened in 1977. By then I was 16, and was in Wales with a cousin on my first independent holiday trip. Travelling round various youth hostels, on one of our journeys we stopped at the Centre for Alternative Technology in Machynlleth, which had only just opened. The technologies immediately grabbed my interest. There were demonstrations of solar panels, wind turbines and hydroelectric power, and information on ecological processes and techniques such as composting and recycling. These are all commonplace now, but they were pioneering in 1977. Again, I had found something that sparked my curiosity. It reinforced my earlier interest in engineering and the wider environment, and also prompted me to join Friends of the Earth. I spent time wondering why we live in such a wasteful society, and why more people weren't involved in the effort towards sustainable living. I read Lovelock's work on the Gaia theory, the proposition that the whole earth is one enormous living organism, in the sense that all processes are interlocked and the earth is constantly moving in interwoven cycles.

Combining my early love of rocks and fossils with my growing interest in engineering, I went on to study geology, technology and engineering. My first job was actually in the oil industry working offshore. It took me round the world. I saw West Africa and the Middle East. Looking back, it wasn't perhaps the best career for a budding environmentalist, but then again it grounded me in reality, provided me with industrial experience and taught me self-reliance. You have to find a very practical and anchored way of living in order to survive offshore, without all the available land-based support systems. It gave me my first taste of finding my own solutions, of repairing and even making my own equipment when things broke. Offshore life also brings a huge amount of work, downtime and rest, with 12 hour shifts and tours of 3–4 weeks working 7 days solid. It's very much a case of work hard, play hard. You can only spend so much time eating, fishing and watching videos, but the rest of the expanse of spare time was, for me, spent in reading and contemplation. It was then I decided that one day I would build my own house and I spent time considering designs and specifications. This would, one day, become another opportunity for practical problem-solving, and a new perspective on the environmental issues that preoccupied me.

Before that could happen however, I had to change career. The oil exploration industry went into recession, and I moved into energy management and traditional building services. I spent time in both the private and public sectors. This culminated in a job working for a building services commissioning specialist company, Commtech, where I headed up the energy division, often focusing on building energy audits and also working as a commissioning manager on some large projects. I stayed there until I founded my own business, Energy Solutions Associates, in 2007. In my spare time, I made my dreams a reality and built my own house in 2000. What finally made me decide to do it was a family holiday to Canada in 1998. In Canada, I noticed how many good-quality self-build houses there were. It could clearly be done.

So, when we got back to the UK, I started looking round for a suitable site. I found an old bungalow in Suffolk, which was quite literally falling down, and so I knocked it down and built my house on its footprint. Sticking to my environmental principles, I was able to reuse a significant amount of the footings and saved many of the materials. I recovered a lot of the timbers and bricks. Any material that would otherwise have gone to landfill was hand separated to be crushed and used as recycled aggregate for the building over-site and a local farmer's track. In short, I tried to be as low impact as possible, although in reality there is no such thing at present as a carbon neutral build. It's more a case of minimising impact and trying to be as low carbon and efficient as realistically possible. Still, minimising the impact was made both in an economic and an environmental sense. By utilising what was there before and using off-site construction techniques, it came in on time and on budget. All extra timber came from stewarded forested sources, and the house was also designed to be way above the Building Regulations (2000) in terms of thermal efficiency. It was also built to have accessible services and easier maintenance, where you can get to the pipe and cable runs. That means that I can alter things when I want to and enables me to upgrade technology as it develops and becomes more economic. For example, the mechanical thermostats have now been replaced with electronic ones, enabling all the rooms to have individual time and temperature control via under-floor heating, powered by a heat pump. Overall, we have roughly half the fuel bills of a traditional house of the same size. (See Chapter 15 for the details and the whole story.)

As an engineer and environmentalist, I try to practise what I preach. I still drive a car – I'm a realist, not a fundamentalist, and I live in a very rural area so I need it – but it has a small one-litre engine, so I minimise my impact and save money at the same time. For me, it's all about common sense. Why waste money and resources?

These real-life experiences are the credentials I bring to this book, along with my years in building services engineering. In my professional life, I haven't just worked to deliver energy management design and audits: I'm also a member of the regional committee of CIBSE and am regularly employed to give high-level training to my fellow engineers. I chair the Eastern region SummitSkills group (our sector skills council) as well as Sustainable Built Environment East, a unique group comprising all the major professional and trade associations in the building sector, in the east of England.

All these years later, however, I'm still influenced by the ideas and experiences of my youth, in particular James Lovelock on Gaia theory. This has influenced and formed that basis of my beliefs as an environmentalist. I'm both an engineer who became an environmentalist and an environmentalist who became an engineer. I still hold that by upsetting the balance of the earth, we are threatening our survival as a species. Everything we do has an impact. In the second decade of 21st century, our overriding issues are about world population growth, and the almost inevitable end of fossil fuels en masse, which we are set to see in our lifetime. There is pressure on water resources and food production, and also on mineral resources. Because of all these happenings, we are having a direct impact on our climate. As humans, we see things in short periods, because we are only seeing things across our lifetime. But if we take a holistic view, and examine what has happened to our planet over millennia, we find proven scientific facts from peer-reviewed data. Extremes in temperatures and climatic change have happened before, but always through natural processes. Modern humans originated in Africa some 195,000 years ago and then migrated to the rest of the world starting around 60,000 years ago (National Geographic 2007); but our real significant impact on the planet only started around 150 years ago, as the technologies of the industrial revolution were exported around the world. Remember, the original industrial revolution was a British-led affair, and as such its impact was relatively limited. As the rest of the world industrialised, the impact began to accumulate. In the future, it should be noted, full industrialisation in China and India, given their massive populations and potential for economic growth, they could yet make our impact to date appear minor in comparison. There are so many studies and graphs published showing varying degrees of evidence for climate change and the links with our industrial activities. For me, the following graphs say it all.

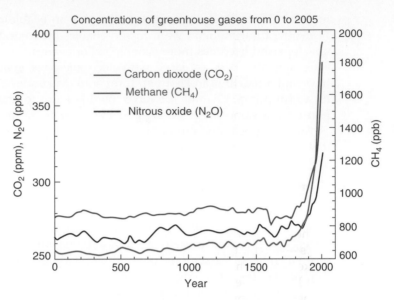

Figure 1 Graph showing concentrations of greenhouse gases from year 0 to 2005 (IPCC 2007)

We can see the evidence for the dramatic rise in greenhouse gases in such a short period, and I am convinced it is still accelerating, as year on year measurements continue to see the levels rise (NOAA 2012). It's no coincidence that the rapid rise in the production and use of fossil fuels between 1800 and the present, shows a startling correlation to the rapid rise of greenhouse gases over the same time span.

Because of this, within my current professional role as a building services, energy and sustainability specialist, I see sustainability as fundamental to everything I do. I see my role as not just about making things work, but about influencing others to see why we need to do things in a certain way. The current inequality of resourcing is inexcusable. People often mistakenly believe that it's about saving the planet. It's not. The planet is fine, and will survive perfectly without us, as it's always done. It's us, the human race, that we've got to save. If we look back through the historical record etched into our rocks and soils, we can see all through geological time a series of mass extinctions. The most famous was the dinosaurs, but there were also others, through various climatic changes. Some were caused by super-volcanoes, extra-terrestrial impacts and some by shifts in the earth's axis and rotation or sea-level rises. Whatever the causes, there have always been rapid and profound changes. We are going through a profound – if largely unremarked upon – change at the moment. When you listen to scientists and naturalists, you hear the warning that we are going through one of those phases of mass extinctions of species. This hasn't happened because of a volcano. This is happening because of us. In his television series 'State of the planet' Sir David Attenborough examined the main causes of damage

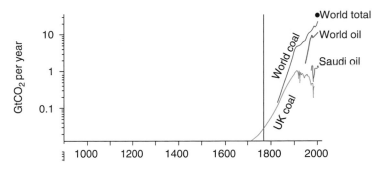

Figure 2 The Industrial Revolution – Fossil fuel production from 1800 to 2000 (Mackay 2008)

to the natural world produced by humans, pointing out that up to 50% of the species on this planet could disappear during this century unless we make radical changes to the way we use resources. If we fail, he argues, we will have made a radical and irrevocable change to all future life on this planet. Attenborough also made a BBC Horizon Special 'How Many People Can Live on Planet Earth?' This is compelling viewing and I would recommend it to everyone.

This book cannot attempt to solve or fully discuss these global issues, but we do need to bear in mind that we can't do what we do in our work in isolation. This book is about what we *can* do in the building services and the facilities management professional spheres. Ultimately, it doesn't matter if you believe in climate change or not, because it makes business, economic and practical sense to run buildings as efficiently as possible. What makes people tick? For me, it's a passionate belief in getting things right and protecting the environment as much as a pragmatist can, but I recognise that not everyone is as passionate as me on these issues. Nonetheless, we all want to save money and resources. If I can help you save money, and it also benefits the environment, then who's going to argue with that?

References

IPCC (2007), Graph showing Concentrations of Greenhouse Gases from year 0 to 2005 IPCC Fourth Assessment Report, Climate Change 2007 (AR4) http://www.ipcc.ch/graphics/ar4-wg1/jpg/faq-2-1-fig-1.jpg (accessed 13.8.2012)

MacKay, David JC (2008) 'Without hot air' UIT Cambridge *National Geographic*, January 2007

NOAA (2012) National Oceanic and Atmospheric Administration; Monthly data Atmospheric CO_2 at Mauna Loa Observatory, March 1958 – February 2012 ftp://ftp.cmdl.noaa.gov/ccg/co2/trends/co2_mm_mlo.txt (accessed 13.8.2012)

'State of the planet' (2004) DVD – PAL BBC

Highly recommended viewing

BBC Horizon Special 'How Many People Can Live on Planet Earth?' (2009) http://topdocumentaryfilms.com/how-many-people-can-live-on-planet-earth/ (accessed 13.8.2012)

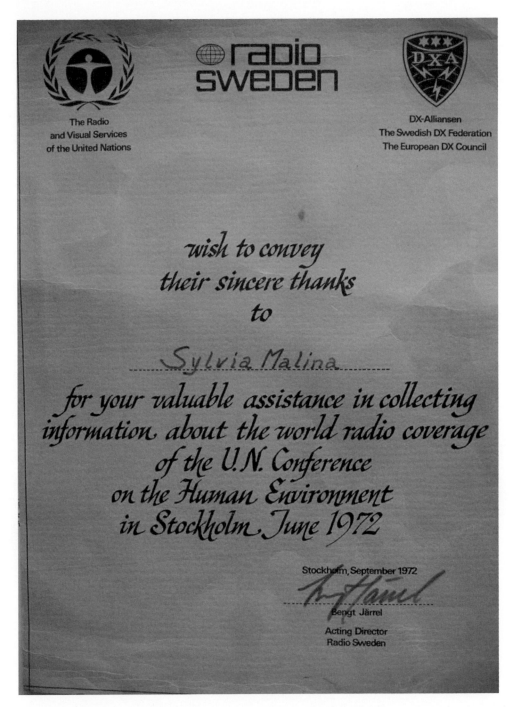

Figure A The first international conference on the environment, the Stockholm Conference 1972 – Sylvia Malina's certificate

Figure B My early interest in renewable energy took me to Burgar Hill, Orkney, in 1986, the site of the UK's largest experimental wind turbine at that time

Figure C This cartoon has followed me from 1979 – student days – and acts as a light-hearted reminder of what happens if we get sustainability wrong (Reproduced by kind permission of 'Brick' – www.brickbats.co.uk) 'To Let – Suit Ambitious Amoeba'

About the book

What is this book about and who is it for?

Anyone who is interested in bridging the gap between the wider sustainability debate and the technical issues of building services engineering, in delivering a sustainable built environment should read this book. It will tell you what you need to know to fulfil current legal requirements, but much more than this it will make you think about the wider long-term issues and how, in order to prepare for future challenges, we need to have an understanding of the 'bigger picture'.

My aim is to highlight the current issues around sustainability and energy use, looking at what is going wrong in the present system and suggesting some potential solutions. I also want to encourage debate among professionals in the field, people who understand the day-to-day realities of working in the construction and building services industry or of managing a building. It is not designed to be a textbook, as you can get this from the excellent resources provided by BSRIA, B&ESA, CIBSE, ECA, etc.

This book is very much a personal view, deriving from many years of experience within the industry and an even longer time being genuinely fascinated by the world and how it works, from the interconnected workings of the natural world to the vagaries of the political system and financial markets. I want people to read this book and believe that energy efficiency is both possible and desirable and that it will not happen by technological advance alone; it will require personal understanding and shared responsibility. Education and awareness is a vital component.

Getting it right will make organisations more efficient and save them money. Getting it wrong is not an option.

After all, if you can save money, lessen your environmental impact and address the crucial issues of climate change and resource sustainability, who is going to argue with that?

About the author

Mike Malina is the founder and director of Energy Solutions Associates, which is a building services engineering practice that works in the field of sustainable engineering, energy management and training.

He has 30 years related experience, working at the start of his career in the offshore oil industry, then working in both the public and private sectors in related buildings and services sector. He is the principal trainer for the Building & Engineering Services Association (formally HVCA), Building Regulations Competent Persons Certification scheme for commercial and domestic HVACR work.

Over many years Mike has conducted hundreds of building energy audits and has never failed to find ways to save energy. In 2010 he won the HVR Consultant of the Year award and in 2011 the Innovation and Sustainability Outstanding Contribution to the Industry award.

In a voluntary capacity he serves as the chair of SummitSkills East and chair of Sustainable Built Environment East. He is also a member of the CIBSE Eastern Region Committee.

Dedication

To all my family – past, present and future.

Acknowledgements

This book is the result of the many hours of dialogue with many people throughout the building services industry and beyond. My thanks to all of them – you know who you are. Indeed many have helped, but it would be somewhat ungracious not to acknowledge a few people by name. In no particular order, I'd like to thank, Terry Wyatt, Bob Blake, Roger Clark, Cath Hassell, Roger Carlin, Nick Ward, Ant Wilson, Rod Pettigrew, Karen Fletcher, Uly Ma, Richard Brown, Ian Ellis, Dean Clackett, Dave Mervin and Lucien Dop.

I'd like to particularly thank my wife Sue for her support and encouragement throughout this project and to Caroline Collier for her patience and keeping me on track in the writing and production of this book.

Mike Malina, March 2012

'In the realm of ideas everything depends on enthusiasm;
In the real world, all rests on perseverance.'
Johann Wolfgang von Goethe
Writer, biologist, theoretical physicist and polymath. 1749–1832

Glossary of abbreviations

AA	Automobile Association
ASHRAE	American Society of Heating, Refrigerating and Air-Conditioning Engineers
AHU	air handling unit
BAP	biodiversity action plan
B&ES	Building and Engineering Services Association
BIFM	British Institute of Facilities Management
BIS	(Department for) Business, Innovation and Skills
BRE	Building Research Establishment
BREEAM	Building Research Establishment Environmental Assessment Method
BSRIA	Building Services Research and Information Association
CCC	Committee on Climate Change
CDM	Construction (Design and Management) Regulations 2007
CIBSE	Chartered Institute of Building Services Engineers
CSA	Commissioning Specialists Association
CSR	corporate social responsibility
DECC	Department of Energy & Climate Change
DEFRA	Department for Environment, Food and Rural Affairs
DTI	Department of Trade and Industry (now called BIS)
ECA	Electrical Contractors Association
EPBD	Energy Performance of Buildings Directive
EU	European Union
FCU	fan coil unit
FM	facilities manager/ment
HVAC	heating, ventilation, air conditioning
HVCA	Heating and Ventilating Contractors Association (Now the B&ES)
ICE	Institute of Civil Engineers
IPCC	Intergovernmental Panel on Climate Change
LCT	low carbon technology
LEED	Leadership in Energy and Environmental Design
NGO	non-governmental organisation
PII	Partners in Innovation scheme – a DTI scheme
RIBA	Royal Institute of British Architects
RICS	Royal Institute of Chartered Surveyors
ROI	return on investment
SBEE	Sustainable Built Environment East

SDC	Sustainable Development Commission
TCPA	Town and Country Planning Association
UKAS	UK Accreditation Service
UNFCC	UN Framework Convention on Climate Change

Introduction

This publication is not just another technical tome about sustainable building design; this is a book about reducing costs and saving money in the long term. In particular, it's about reducing the whole life costs of a building, and therefore reducing the overall cost of ownership. The bonus, of course, is a reduced carbon impact and greater energy efficiency.

Building services are not the first things that we notice when we look at a building. What most people see is an impressive facade, or perhaps the shape of an architecturally interesting roof. It's easy to forget that all that is merely a shell without the components which make a building fit for function and occupation. A building is useless, even as a warehouse, without building services such as heating, ventilation, air handling, light and power. These services are an integral and vital part of the building. They are also central to its energy-efficient operation.

Early modelling and design (BIM)

It makes sense that these services are considered from the earliest possible stage in the design. You need good-quality engineering if you want an efficient and optimally performing building, and it has to be planned for from the outset. Too often this is where things go wrong and important opportunities for saving energy and reducing carbon emissions are missed. Buildings represent 40% of the world's energy consumption (IEA 2002) and energy accounts for 40% of the actual building operation cost (Carbon Trust 2012). These figures can vary, but it will certainly be a significant sum. If services are developed as an afterthought, which they often are, the building will almost inevitably cost more to run, both financially and environmentally.

Encouraging the design team to work more closely together is challenging. However, building information modelling (BIM) is a software tool that is helping to encourage greater cooperation in construction teams. It can also be seen as a tool to encourage and promote a more sustainable and cost-effective way to deliver sustainable buildings. BIM applies software systems to evaluate and maximise the efficiency of the project construction. It works as the foundation for collaboration in design and construction, to ensure that

Delivering Sustainable Buildings: an industry insider's view, First Edition. Mike Malina.
© 2013 Mike Malina. Published 2013 by Blackwell Publishing Ltd.

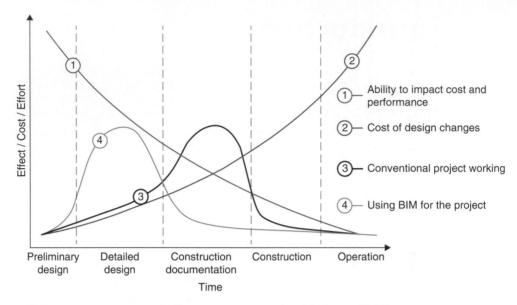

Figure A Design costs, impacts and influence on project timeline (MacLeamy 2005)

project stakeholders including client, architects, consultants, contractors and facilities managers have access to a collective system that includes all the details of the projects design, specification, materials, project plan and costs. As part of the process BIM will produce a 3D model of the project.

The industry will see BIM become more commonplace over the next few years and this has already been identified by the UK government as an important part of its construction strategy, which it published in May 2011 (GCS 2011). The government identifies a 20% improvement in efficiency of construction using BIM, and has stated that it intends this method to be phased in for all government contracts by 2016.

Figure A can help us think about project timings and their impact on design and construction costs (MacLeamy 2005). The project timeline runs across the horizontal axis, while effort, cost and effect are shown on the vertical axis. If services are considered at the start, you get the maximum impact and effect. As the process goes on, however, it becomes more difficult to influence the changes without problems and excessive costs.

The first line shows the positive results of designing in services early, with the ability to gain maximum influence on cost and performance. The second line illustrates the cost of changes. We can see that costs rise significantly, as other factors become set in stone and the physical reality of the building makes alterations more challenging. If you are trying to save money, late changes will work against you.

Line three represents the traditional decision-making process, representing the majority of projects as they have been conducted up until recently. As we can see, this does not happen as early as would be optimal, so costs can spiral and everything happens late within the construction process.

In contrast, line four shows a robust design process, using BIM, which models everything before the start of the project. In this model, building services have significant early influence, meaning that both effort and reward come early in the process and the team can head off the majority of problems from the start. The later things happen, on the other hand, the bigger the impact on costs and potential delays. The same pattern also holds true for refurbishment projects. The more thought and effort that is put into early planning, the better. With the process of BIM, both the costs of the project will be lower than conventional methods as well as the whole quality of the project's delivery.

Costs tend to override everything in construction, particularly in hard economic times, but the mindset of the industry needs to change to the point where it realises that just a little bit more thought and early intervention will pay massive dividends for the future. Savings on operating costs and handover will actually have the biggest impact of all, since the energy consumed over the lifetime of a building is phenomenal. The other factor that will become an increasingly important issue for the future is embodied energy. This will probably be included in the total lifecycle calculations and added to the operational energy measurement of the building.

So, what constrains progress? Of course, a major issue is that often when a client wants a building, it's not actually for their own use. The client may be a developer. Therefore, they have no obvious incentive to consider whole life costs. Legislation in this area is developing however, to tighten up the process and address this issue. There is also the growing role of corporate social responsibility (CSR). As a society, we have to be more realistic about the sustainability of the world we are creating and influencing. If concerns about the future of the planet are not sufficient motivation, the other factor that developers should realise is that they will get a better return on investment (ROI) on a building that will operate efficiently over its lifetime. Lower operating costs equate to significant added value for a building's occupants. In a buyers' market, those buildings that can demonstrate a higher level of energy-efficient performance will be more attractive. Some of the larger developers and estate agents are starting to realise that this is in their interest, particularly at the time of writing when there is a glut of office space, with buildings lying empty. Low-efficiency buildings can't be marketed, whereas higher-efficiency buildings can. If a client is planning to occupy a building themselves, there is obviously a clear incentive for them to specify a building which will run efficiently.

Despite all this, until very recently architects tended to have little understanding of building services. In this situation, it almost inevitably becomes a bolt-on. It is vital for architects to talk to consulting engineers and to the designers of mechanical and electrical (M&E) building services. Unfortunately, because of the way the construction process has traditionally worked, the M&E engineers sometimes arrive on site with a bare minimum of information, and have to design services in a short time. They will then have to do their best to fit these into an already constructed environment. This means that you can end up with a ridiculous situation, with the engineers trying to work with drawings that don't bear any relation to what exists on site.

For instance, the ductwork, in the drawings, looks as though it could be straight, but when the engineer arrives, they find it actually has to bend around numerous architectural features. This extra strain on the system will make the ductwork less efficient, and there will be a knock-on effect. The engineers will have to redesign the services to cope with all this extra resistance. This might mean bigger fans or motors in air handling units, which consume more energy and therefore cost more to run. This is a simple example of how, because the services were not thought out early, a building can be delayed and become more expensive to operate. Going back to Figure A, having to change things late means that the engineers have less influence, and the effect from their effort goes down. This needs to be addressed, because the majority of construction projects currently suffer from problems of this type.

Changing the current modus operandi in the face of the real financial and time constraints, which are inherent in the industry, will take a lot of effort and education. Just as politicians look no further than re-election, some construction professionals look no further than the next job. The industry needs to develop a longer-term vision – as, indeed, does society as a whole. The construction industry needs to nurture a sense of pride in its work and outputs, to remember that it is creating buildings which will stand for generations. These buildings do not only need to be aesthetically pleasing, they need to work efficiently. Despite this, the notion of checking that everything is working and that the new owners are happy beyond the handover is very new. Contractors, including M&E engineers, are not usually looking beyond installation. Nonetheless, if the industry is to deliver the products society needs to meet the challenges of the 21st century, it is vital to look at how things are performing and to find out whether the people using the building are satisfied with it. Construction professionals have to create a usable and integrated environment that will work for its occupants. Ultimately, a functional, calibrated building is less costly.

Not only that, but studies have shown (Heerwagen 2000) that workers perform better in well-functioning buildings. A pleasant working environment therefore creates value in more ways than one. People are more productive if their premises are well commissioned, maintained and operated.

Currently, if we talk to people actually using buildings, are they happy? In my experience as a consultant, I have found that when asked this question, a lot of people say no. They often feel that they have no control over their environment.

There is currently a debate about how much control of their environment people should have. Some engineers and designers would like to automate everything. They like the idea of taking decisions out of people's hands and relying on technology for regulation, but this may not always work. In practice, people can feel alienated if they feel they have no control over their surroundings. But, at the other extreme, too much control in inexperienced hands can throw the system out.

Historically, building service engineers have not liked the idea of ordinary people interfering with building controls. There have been instances of placebo controls being installed; dummy thermostats, for example, that look and feel like the real thing (in fact, they are the real thing – they're just not connected) make people think they have that all-important control. There has been some evidence to show that because people can adjust these fake controls they feel more in control of their environment.

Another instance of this approach to occupant perception is an occasion where people in an office requested more natural daylight-coloured lighting. As there was no budget for it, the building's facilities team cleaned the light fittings over a weekend instead. Coming in on Monday morning, occupants had the perception that the light fittings had been changed, and were satisfied.

With care, psychology can be used to impact on how people interact with, and understand, their environment, but it is certainly a dangerous game. It's generally far better to explain circumstances honestly. If building users were educated better, they would come to understand what is physically possible.

Creating a satisfactory environment is also about good housekeeping. That means that commissioning and maintenance are not the only issues engineers have to think about. They also need to make a big effort to explain to users what they can achieve in local operation. If an environment is operated incorrectly, users can effectively destroy its set-up. All too often, people fiddle with controls that they don't know how to use. For instance, there might be a unit on the wall for the air conditioning, or a handheld set of controls. People often don't fully understand these controls. They might use them to raise and lower the temperature, but do they ever reset it? Too often, common sense deserts building users if they find they can't control the environment effectively with the controls. What they often do is physically open the windows. This is tantamount to throwing money down the drain.

To achieve optimum results, there has to be a compromise to achieve the best of an automated system while still leaving occupants feeling that they are involved. Success here requires a culture of trust between the designer, the building services engineer and the building users. I believe that the only way for this to be done is through good design, operation and commissioning, supported by effective training and handover. If all these are carried out correctly, the building will operate efficiently and people will feel that they have more ownership of their own environments. This is a debate that needs to be thoroughly explored within the building services industry.

There is also the issue that people today generally have very high expectations; sometimes these expectations are too high. After all, technology can only do so much and people need to understand what it can and can't do. Air conditioning, for example, is a catch-all term. A lot of buildings have localised comfort cooling. This is termed 'air conditioning', so people expect almost magical control over their environment, but often all it does is physically cool the air. It's just a localised intervention, and people have little control over it.

Also, different people have different perceptions. Some people feel the cold more than others, and people with different perceptions of heat share the same offices up and down the country. People also often forget to dress for the weather, working in clothes that defy common sense. People come to work in sleeveless shirts in winter and expect to be kept warm, and then do the opposite in summer. Due to design issues and energy consumption, this is not always physically possible.

If buildings are not commissioned and maintained properly they will have 'starved' systems. In effect, one end of the building is hot, while one end is cold and what you end up with is system that is not balanced properly. Where it can't reach a comfortable temperature, people bring in electric heaters. This of course creates more energy consumption and puts a strain on the electrical load, raising numerous other safety issues. While this is going on, at the other end of the building where all the resources are being used, the system overheats and people open windows. This situation is all too common, and can even happen with a brand new system. The reason for this failure is poor commissioning. This could be alleviated by a good building services explanation and handover, and investing in the processes of continuous commissioning.

Returning to the diagram in Figure A, which shows effect and effort, you could equally apply a compressed version of this illustration to the commissioning and handover. Once again, you could have a fantastic building that's designed to the highest standards, but unless it's commissioned and handed over to the users properly, all the construction effort is a waste of time.

A useful analogy is to think of a car. It might look nice and run OK, but if the engine is not tuned, it won't run properly. In the same way, you need to think about fine-tuning building services early in the handover process. Using another aspect of the car analogy, when you learn to drive, you learn how to use the gears. You learn not ot push your revs over a certain level and not to go hard on the gas. It's only later that you develop bad habits. In the first place, you aim to pass your test and do everything perfectly.

Unfortunately, most building users have never even been taught to 'drive' their building in the first place. But there is a sense in which you should learn to drive your building. If you drive efficiently, if you use the gears appropriately, you will spend less money and there will be less wear and tear. Well maintained and calibrated buildings are also more efficient, because they don't waste heat, light and energy in air movement. It places less strain on a building's resources, and is also good for maintenance.

Unfortunately, as with driving, people pick up bad habits. They get lazy. We all take things for granted, and it's all too easy to come into work in the morning and just flick all the switches on. People automatically turn the heating on or up without really thinking, or leave in the evening without switching everything off. Computers and lights are routinely left on all night. Even if the lights are on a timer, that timer may well not have been altered when the clocks changed. The consequences are just the same as if we neglect a car, and forget to check the air pressure, oil and water. Of course, not all occupants need to do all the checks. However, like the passengers in a car, they have a vested interest in the building working to its optimum performance. A wealthy client is perhaps more like the passenger in a chauffeur-driven vehicle. They put their trust in the driver, just as the client puts trust in the building services engineer. Both roles need to be filled with competent and conscientious members of staff for reasons of both safety and finance.

When going to a car showroom to select a car, we increasingly look for efficiency. We also consider what we will use the car for. Do we need to be able to move a piano, or just people? Does the car need to be prestigious, or do we just want a run-around? There are a whole range of questions. Once again, the same applies to building services. What is the building's function and what want do we want it to do?

The engineering has to meet the client's expectations. Clients don't have technical knowledge by and large; they just say 'I want a heating system'. It's the engineers who have to select the right technology. If the engineers don't specify correctly and understand the overall concept of the building, things will go wrong. Some technologies are better suited to some designs rather than others. It's not possible to just put any system into any design. There are common patterns – just as all cars have four wheels, all buildings need heat, light and ventilation – and there are common systems to use, but we have to select the right technology. How does it all fit together in the design? What is it to be used for? How much control does the client want, and how much needs to be automatic? If the engineer gets this right and meets the client's needs, then it might cost more to start with but it will decrease the cost of ownership over the life of the building. It's simply about getting the right technology in the right place for the right operation.

Ultimately, whole life costs of a building can be considered with regard to the whole project. If construction is planned well, it will pay dividends. It always pays back. Traditional methods of construction deal with building services on an ad hoc basis, but integration brings real benefits over time.

Technologies and applications

Every building is different. This doesn't just apply to appearance and design, but also to geographical location. Is the building urban or rural? Location often defines what type of technologies can be used; for example, in a semi-rural or rural environment, it is possible to use more natural ventilation. In a city centre, constraints come from conditions such as air pollution and noise pollution. Therefore, a building is likely to have to use more mechanically controlled ventilation.

The other factor is orientation. How is a building affected by natural processes such as the sun? How much natural heat will build up? Natural heat can be both good and bad. Solar gain is beneficial in the winter, but an engineer will want to restrict it in summer. Also, services must take account of the nature of the building envelope. The fabric, even the colour, must be considered. Reflective material is desirable in warmer climes, and with changing weather patterns and climate, more areas will need to adopt these measures. Insulation can be both for heating and cooling. Sometimes we need to keep heat out, and people often forget this. There are situations where we need to stop solar gain, which can go through walls as well as windows if they're not insulated. There are often good examples of this problem in factory units and warehouses. A metal-clad exterior can become like a frying pan, picking up the heat. Without reflective material, the internal environment will become oppressive and the occupants will have to disperse heat. A lot of these problems should be addressed through the Building Regulations. See Chapter 3 for a discussion of the Building Regulations Part L.

When it comes to ventilation, we should try to use natural methods wherever possible. If the cooling and air movement can be free, it would seem foolish in terms of both expenditure and the environment not to use it if possible. Despite this, there are many examples of buildings that could be adapted to use free cooling that are not, and remain reliant on mechanical systems. The reasons for this are cultural, and need to be addressed at a societal level. People expect technological solutions for everything. Therefore, some things tend to get specified automatically, and there is a tendency to over specify. This is not done to intentionally disadvantage the customer, but because engineers become conditioned to do it. Lighting also needs to be considered with regard to orientation, as natural daylight will obviously be affected by this. Where there is free light, we should take advantage of it. Nevertheless, many buildings have their lights on all day for no reason. As a society, we have stopped taking account of natural daylight. This is both cultural and habitual. We could do it either manually or through technology – a simple daylight sensor could switch the lights off for us.

The energy hierarchy diagram (Figure B) is absolutely key to effective, low carbon services design. The most important step towards energy efficiency in

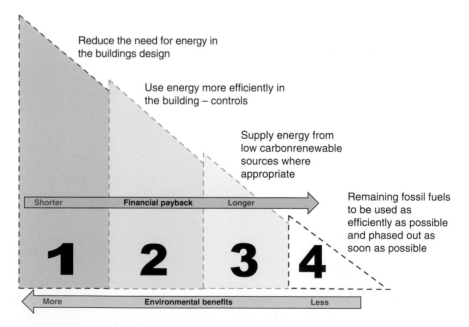

Figure B The energy hierarchy – What delivers a low carbon energy-efficient building?

building services is to make this hierarchy the underpinning strategy behind every design and operation for the future. It's a simple but extremely effective strategy. It's also important to remember that achieving a good plan based on this hierarchy will also improve the cost–benefit analysis on projects. Too often people think that 'green' building is expensive, but energy efficiency and cost savings are actually coterminous.

The hierarchy shows us that it is important to reduce the need for energy in the building's design in the first place. This stage will include things such as making sure that the building's lifecycle is taken into account. It's important to examine the energy flow in the entire lifecycle of all materials that go into the building, so first there is a need look at the external envelope of the building. It's important to conserve the energy, so why design something that will be naturally wasteful? Having studied the building envelope first, the designer and engineer then need to ensure that all the systems are integrated so that they naturally use the least energy possible. At the same time, it's important to look at the carbon impact and usage of the materials involved.

In the first stage of the design process, therefore, the designer has to look at the embodied energy, which refers to the amount of carbon used to manufacture and transport the materials and the energy input within the process. This should be the first priority in any building design. Secondly, it's important to use energy more efficiently in the building once it's being operated. This can be significantly affected by the early design process. The key point in this second part of the process is the issue of systems integration and controls. This involves utilising building technology and controls to monitor and operate the building services, and to make sure that all the building services doing different things are integrated, not competing against each other. It means working for the benefit of the operation of the systems themselves. This will be covered in more detail in the building controls Chapter 10.

This hierarchy means that steps one and two can be taken into account from the start of the design process. Only then do we need to consider the third element of the hierarchy, and to start looking at the supply of energy from renewable and low carbon technologies, because this can only sensibly be done once steps one and two have been undertaken. At the time of writing however, there is an increasing trend, within society, with building end users and with designers, for people to want to start the energy efficiency process with step number three. This is because there is a growing awareness of the need to source energy from renewable sources, and people want to help this process along. People want to be green. Therefore, many are going out and spending large capital sums on renewable and low carbon technologies, as they think it will help mitigate climate change.

Unfortunately, this trend can actually be detrimental to carbon consumption if it means that this aspect is focused on to the exclusion of working on steps one and two of the energy hierarchy (Figure B). The reduction in embodied energy from alternative technologies can be really insignificant in comparison to the savings that would be achieved by getting steps one and two correct. There would be a much better return on investment, in terms of both carbon reduction and financial savings, if these steps were taken to reduce the need for the energy in the first place.

When nuclear power first came along, the catchphrase was that it would be able to provide limitless energy that became 'too cheap to meter'. We now know that this was totally wrong. It's about the most expensive form of energy going, factoring in the issue of nuclear waste that always seems to be ignored in these equations. Despite this, similar expectations are forming around renewable energy sources – once again, as a society, we expect it to produce limitless cheap energy. Sadly, it's not that simple. The technologies involved are very expensive, though they will reduce in price as they become more established.

Speaking as a practitioner, I believe that there are far better things to do to save energy. By starting with steps one and two of the energy hierarchy, we can reduce our overall need for energy and increase our control over its use. People feel good about installing a solar photovoltaic panel, but it takes more than ten years to pay back. If, by contrast, we reduce energy need, by installing insulation, for example, the payback is comparatively very quick indeed.

The energy hierarchy provides a total correlation between its priorities and the likely level of financial payback – adopting step one will provide the greatest dividends. Step four will take far longer, although it will become cost effective in the longer term. Historically, technologies such as solar photovoltaics have had an artificial boost from the feed-in tariff (FIT), a government-led market mechanism that guarantees the owner of the renewable energy system payment for the energy, which is up to four times the value of a unit produced and purchased conventionally. This has been used as a stimulus to prime this market, but as the market develops this will no longer be economically viable. As it is, it is funded by the tax on carbon. As renewables become mainstream, the tariff starts to be reduced. This reduction began in 2012, somewhat controversially, but it was inevitable. I believe that this was the correct thing to do, but the government handled this whole issue very badly and caused a great deal of confusion with the wider population and threw the industry into turmoil.

Ultimately there has to be a balance between technology, deployment and finance. This is a theme that I try to develop throughout this book and raise in more detail in Chapter 4 on finance.

I believe that the future for sustainable buildings will involve the use of the BIM process that makes cooperative working in the construction team more realistic. We will also see a much closer consideration of the building's lifecycle, including its energy use over the long term, and its actual performance compared to the design. At the same time continuous commissioning and maintenance will become much more important.

These principles will be the bedrock of a holistic approach that offers the best possibility of achieving low carbon and sustainable buildings.

References

Carbon Trust (2012) – Buildings Policy http://www.carbontrust.co.uk/policy-legislation/ business-public-sector/pages/building-regulations.aspx (accessed 13.8.2012)

GCS (2011) Government Construction Strategy, Cabinet Office, May 2010

Heerwagen J, (2000) *Green Buildings, Organisational Success, and Occupant Productivity* Published in a special edition of Building Research and Information Vol. 28 (5), 2000:353–367 London, UK

IEA (2002) International Energy Association http://www.iea.org/index_info. asp?id=2401 (accessed 13.8.2012)

MacLeamy (2005) adapted from the 'MacLeamy curve' http://www.msa-ipd.com/ MacleamyCurve.pdf (accessed 13.8.2012) also based on Barrie, Donald S. and Boyd C. Paulson, Jr., *Professional Construction Management*, McGraw-Hill Book Company, 2nd edn, 1984

1 Sustainability in the wider context

1 Making the right choices – the sustainability dilemma

Exactly how do we make the right, sustainable choices? There are so many competing facts and figures, and a lot of conflicting information from well-meaning campaigners, business, government, non-governmental organisations (NGOs) and trade bodies. Everyone has their own agenda and opinions.

There is a wealth of information from industry as well as legislation and standards, and a lot of this creates conflict, which reflects opposing interests. In any process, in business or buildings, there are differing views and product loyalties, but in the field of sustainability the problem seems to be particularly acute. How do we cut through this? How do we create a transparent system to make sure that everyone gets the right technologies? There are so many claims for products, which can be oversold and mis-sold. Therefore we need a level playing field involving testing, transparency and accountability.

Objectivity is the key

I would argue that the only solution is to be as objective as possible. I would always approach every claim – and every adjustment to conventional technology such as proposed enhancements and renewable technology developments – as the ultimate sceptic.

I work on the basis that you always have to ask the question: does it do what it says on the tin? Just because the product literature says it does something, it doesn't mean it does. Even when it does do what it says on the tin, is it the right application for the task in hand? How is it going to be used and, of course, we must ask: what is its true impact throughout the product's lifecycle and how will it affect and influence the wider project or building? (This goes back to the implementation of the hierarchy of energy, as referenced in the introduction and throughout this book.)

So the key is to be objective. What I believe is lacking are national standards that would truly test every new sustainable product or claim. I think that, as an industry and a society, we are too trusting, and we often like to believe that things are the best thing since sliced bread. A good sales person can exert enough influence for the wrong decision to be made, and it may

Delivering Sustainable Buildings: an industry insider's view, First Edition. Mike Malina.
© 2013 Mike Malina. Published 2013 by Blackwell Publishing Ltd.

be only years later that the buyer, specifier or user finds out that the technology doesn't live up to expectations.

An example would be large utility companies who, at the time of writing, are in the process of setting up significant installation businesses for renewable and low carbon technologies, as they see this as a major market opportunity. The big question is whether this will encourage the tendency for sales people to get carried away with sales targets. As more grants are made available for funding, the take-up of renewable and low carbon technology in the UK, we have to ask how that might influence the selling process. How often have you heard a sales person admit that this isn't right for you and thereby not making a sale? This will be a crucial point, in that there need to be very responsible business attitudes, so internal systems of these large companies need to guard against mis-selling. As an industry, we've got to guard against risking our good reputation with potentially false claims or poor standards, like those associated with the double-glazing industry's reputation of the 1970s and 1980s.

Legislation and industry and government action are required to police the markets and give people the correct information. If the industry is left to function as a free market, poor products will eventually fall out of the system, but this will only work to a small degree. And what will be the cost to the consumer as this process takes place? Surely it's better to get this right from the start? It's always been a difficult situation, because governments want to stay clear of market intervention. And yet, they are still intervening in the market by providing significant stimulus to encourage the take-up of sustainable products – for example, look at the feed-in tariff, or the renewable heat incentive.

It would seem logical for the government to set up national standards for energy-saving and low carbon renewable technologies, to test and rate all these new products. This would give the products more credibility. There could be a common label, independently verified, to promote rigorous national standards, perhaps based on an A to G rating model to measure and benchmark the operational performance. Put simply, A is good – G is not so good. This would create a simple and transparent system, which would allow everyone to judge the relative merits of what a technology does or claims to achieve. This could be done by a range of institutions, perhaps academic bodies, or the National Physical Laboratory. There are also other institutions that have a very good reputation, such as the Building Research Establishment (BRE) or the Building Services Research and Information Association (BSRIA). They could also become part of this scheme, and once this scheme is established, we would then have a baseline to start to judge relative merits of each technology.

At the same time we also need detailed notes and guidance for a product's actual application in non-domestic and domestic buildings. This is because too often at the moment we see a perfectly good technology misused because it has been wrongly specified. For example, using the sun to warm water with solar thermal panels is a good idea in principle, but only if there is a reasonable need for hot water. Putting lots of panels into a small dwelling or office would not be a good application of the technology. So this would need to be part of any national standard involving the use of good application guides. In other words the technology could be A rated for good performance, but be totally wasted if installed in an inappropriate application.

Rigorous standards and enforcement

In the marketplace itself, we need to have a rigorous policing of the standards, and to stamp out bad practices. We have existing legislation that can be enforced by local authority trading standards bodies. These departments need to be significantly enhanced, since they tend to be very small and have limited resources. An example of helpful legislation would be the The Consumer Protection from Unfair Trading Regulations (2008). This superseded the Trade Descriptions Act (1968). This would provide a legal course for claims to be challenged and taken through the courts if necessary. I've always been astounded at how many 'snake oil sellers' there are in the market, an example of which might be magnets on fuel lines or water pipes, which claim to have energy saving properties. I believe these claims to be totally false, because when any of these sales people are challenged to provide robust independent scientifically verified reports, they can never do so. The ultimate question I always pose is if the technology is that good, why aren't manufacturers fitting them as standard? Why aren't the Automobile Association (AA) recommending them for vehicles? In fact, on the contrary, there have been scientifically based reports (Crabb 1997) and a review of tests carried out that showed little value in these claims and dismissed these particular products (Allen 2005; Powell 1998). The Advertising Standards Authority (ASA), upheld complaints from two local authority trading standards departments on misleading statements made by one of these companies (ASA 2002). Yet these companies continue to sell and advocate these products, and people still continue to buy a virtually useless bit of kit. It astounds me when I see these devices fitted in some major companies' plant rooms (Figure 1.1a). The same applies to Electronic 'descalers' (Figure 1.1b) which are also questionable as to their effectiveness.

(a)   (b)

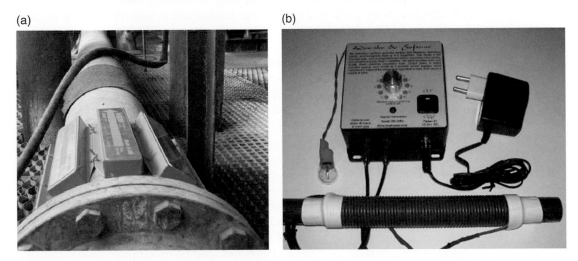

Figure 1.1 Water 'treatment' magnets and 'electronic descaler' – might as well be an ornament

Ultimately we need a strong lead from the government to set up a system of standards for testing and transparent labelling. This should provide all the necessary information to show what actually works and contributes positively to increasing performance and saving energy. This government information could also be extended to the true costs and real-life performance of a whole range of sustainable or low carbon products. It's always nice to feel like you're doing your bit for the environment, which has led to a fashion for what I have termed 'green bling' (Malina 2010). Even Prime Minister David Cameron had a wind turbine fitted to his own house (Guardian 2012), which in reality was nothing more than an expensive ornament. The same applies to photovoltaic (PV) panels.

So many times in my career, I've come across people not understanding that PVs are a developing technology and that at the present time the efficiency and conversion rate of sunlight to electricity is 12–18% at best. Obviously this technology has to start somewhere, and those people that do adopt this early should be made aware of this. This is why the government intervened in the market and created a feed-in tariff (FIT), as it was the only viable way of making it financially economic. Saying that, this could still be regarded as marginal when compared to other technologies and practices, which have a far better energy and environmental performance and provide the best return technologically and financially, very much following the steps of the energy hierarchy methodology. If the FIT was removed or reduced significantly, then this would pull the rug from under the market. So the reality has to be laid out for everyone to see.

There are a number of variants to the way that companies are approaching this market. An example would be the 25 year leasing of domestic or commercial roof space, whereby a company gets the owner of a building to sign an agreement to allow them to place PV panels on the building's roof. The leasing company get the benefit of the FIT, and the building occupier gets the benefit of the free electricity. This is useful from a sustainability point of view, but the offset of the payments for electricity use is far less than the feed-in tariff. That gives you guaranteed money for electricity generated. The owner would get the free electricity, but this is normally priced at 3p per unit, not the 43p offered by the original feed-in tariff prior to its reduction in 2012. The payback was in theory 10 years, but realistically you're not guaranteed the weather pattern that is often used to calculate the projected performance and payback. There are also hidden costs for maintenance: panels will degrade over time, and the inverter devices – which transform the resulting (weather-variable) DC current of the PV panels into alternating current – degrade and will need replacing on average every eight years. They're also costly, being priced at up to £2000, depending on the PV installation size

The other dilemma here is that the companies are leasing these PV panels on a contract signed by the owner of the building, which typically provides for a 25 year lease. So what happens if the owner moves? The contracts are designed so that when the building is sold, the new owner inherits the lease. You would think from a marketing point of view, that most people would agree, and see the benefit for the incentive of free electricity, and more so as prices rise. This may be true for some, but quite a few people would not like to have that feeling of loss of possession. This may create unforeseen problems when the original owner attempts to sell, and this underlines the fact that these things need to be properly thought through. The idea of this type of leasing agreement has been applied in the past to a whole range of major industrial products and plants, and it may well be a financial mechanism for encouraging the take-up of the developing renewable and lower carbon technologies, as many people whether domestic or business owners will not have capital to pay up front for them.

It's the same with the Green Deal: the funding will be made available and all the payback will be funded from the electricity bills as the savings are made. Here again there is a potential pitfall: if you went out tomorrow and brought photovoltaics and then sold the house in three or four years, the panels would be seen as a bonus by some but as a negative by others. It may even be an obstacle to selling, as the contract is with the house rather than the owner. It's a fixed item. We will need a culture change, however, to see this as part of the house, like the newly installed double glazing. Personally, I don't see it as a problem, but it's new and there may be resistance.

I often find myself in a difficult position, as I have wanted to see more deployment of these renewable and low carbon technologies. Nevertheless, in conversation with people who passionately believe in renewables for energy production, I often find myself almost playing devil's advocate. This is because I always come back to the principle and concept of the energy hierarchy. Surely it is better to reduce energy use in the first place rather than to spend more money and waste energy generating even more? Even with sustainable energy, we don't want to get into a culture where we think of electricity as too cheap to meter. This concept is a lesson from history, as this is what many in the nuclear energy industry were forecasting in the 1950s. Nuclear failed to deliver, and this demonstrates the impossibility of truly cost-free energy. We don't want people to think that energy is limitless. There are always going to be some costs, including the energy that goes into manufacturing the PV panels, which are loaded with embodied energy and resources. They also require additional maintenance to the associated infrastructure and can degrade in performance over their operational lifetime.

Throughout history, technologies have crept in and slowly become the standard. It's interesting – can anyone think of a precedent where there has been such a large government-inspired subsidy to encourage technology to this degree? I often wonder, if the government had legislated to put this type of market subsidy and scale of resource into energy conservation, wouldn't it have been a better use of resources to have significantly increased energy conservation? This question is also highlighted by the government's newly created Green Deal. (See Chapters 4 and 13.) This covers renewable and low carbon technology and energy conservation, so this again would be enhanced by the adoption of the energy hierarchy. No one should be allowed grants or subsidy for PV panels without first implementing basic energy conservation. This will hopefully be part of the thrust of the Green Deal.

The 'green deal assessors' (Department of Energy and Climate Change 2010) could be used to deliver such a programme of moving towards a lower carbon society. This would provide a mechanism to truly implement a workable energy hierarchy regime. To this end, it is vital that thorough training is provided to ensure that assessors have the proper skills to interpret a multitude of possibilities and situations. Installation, commissioning, verification of performance monitoring and true financial monitoring will need to be integrated to give an truly accurate picture and give all the facts to create confidence in the development of the low carbon and renewables market of the future. This is discussed further in Chapter 13, which looks at the issue of skills.

Where will our energy come from in the future?

There is a lot of thought going in to the future of energy generation in the UK, as the debate on the transition to a lower carbon economy moves forward. The future of coal, gas and North Sea oil production all have such a major impact, because at present they have such a dominant role in the current economy, and will continue to exert a major influence for the next decade and more. These fuels cannot be switched off or reduced significantly in such a short time. There will be a need to develop a national programme, recognising the importance of energy conservation, coupled with more efficient technological development and deployment. This, together with the large-scale deployment of renewable energy infrastructure, will have to be accelerated if the government targets for carbon reduction are to be achieved. It must also be remembered that the current set of nuclear power stations are coming to the end of their lives. There are ten nuclear power stations across the UK. At present, government planning envisages all but one of the existing nuclear power stations closing by 2023 (BERR 2008). There is a debate developing around what will replace them. This is a whole debate that could fill another book. The government has stated that any new nuclear power stations will be constructed without public subsidy, yet the decommissioning of old reactors and the handling of nuclear waste will be subsidised.

Government subsidies to the nuclear power industry, throughout its history over the last 50 years, have been massive in proportion to the actual value of the energy produced. In a report by the Union of Concerned Scientists (Koplo 2011) a conclusion was made that in some cases it would have cost taxpayers less to simply buy the energy on the open market and give it away to consumers.

The two largest political parties in Britain both see nuclear as part of the UK's energy mix, as well as advocating a massive expansion in low carbon technologies including renewable energy production. Research by the Sustainable Development Commission (SDC) established that even if the UK's existing nuclear capacity were doubled, it would only result in an 8% cut in CO_2 emissions by 2035. The SDC also highlighted many other disadvantages, including long-term waste problems and complications for storage. The cost could be a massive drain on public money, despite the government saying no to a public subsidy.

The design of nuclear power stations is very inflexible. The continuing idea of expanding this type of energy generation could undermine energy efficiency. Finally, there is always the question of international security and potential terrorism. There is a risk attached to the transportation of nuclear materials.

On balance, the SDC concluded that the problems outweighed the advantages of nuclear as a form of energy generation in making a contribution to meeting future carbon reduction and energy needs (SDC 2009).

Public opinion is something else that the government will have to take into account. A recent Ipsos MORI/Cardiff University survey (MORI 2011) found that the British public favoured using renewable sources of energy over and above nuclear power. Solar power was viewed the most popular (88%), followed by wind (82%) and hydroelectric power (76%). By comparison, the popularity of conventional fuel sources were gas (56%), coal (36%), nuclear power (34%) and oil (33%).

Although the present government seems to be pushing ahead with the building of at least four nuclear power stations, Britain and France are to sign an agreement to cooperate on civil nuclear energy, paving the way for the construction of a new generation of power plants in the UK (Guardian 2012). However this pans out with public opinion and environmental campaigners, and the potential for a long planning or public enquiry, this will probably dominate the debate over the next few years.

Figure 1.2 shows the Sizewell nuclear site, which is in my home county of Suffolk. This dumps an enormous amount of waste heat into the sea. Even the new generation of proposed nuclear stations will, after generating electricity, waste the remaining 63% of heat energy in this way.

To deal with the other element of still significant energy generation – coal – the government is also looking at carbon capture and storage. However, I personally see this as tantamount to 'sweeping the carbon under the carpet',

Figure 1.2 Sizewell nuclear site

Figure 1.3 Aerial view of the cooling towers of the Cottam power station, Nottinghamshire
Copyright: Ian Bracegirdle and licensed for reuse under the Creative Commons Licence

Figure 1.4 Government energy policy (credit: Sarah Malina)

as we should be looking to phase out coal and, where possible, look at the cleanest combustion as a transition to the lower carbon economy. Ultimately it's the 'fifth fuel' – energy conservation and efficiency – that should dominate the future, but all governments have yet to fully grasp this as the priority it should be. Figure 1.3 shows an aerial view of the cooling towers of the Cottam power station, Nottinghamshire, where 60% of the energy is also wasted as steam to the atmosphere.

The government has been obsessed with the idea that the lights are going to go out and that the UK needs generating capacity. This has partly fuelled the idea of micro-generation technologies. But if we return to the energy hierarchy, we can see that much of this generation is like pouring water into a leaky bucket. If we're going to plug the holes in the bucket, we need to reduce energy in the first place. I sometimes liken the lack of joined-up government policy on energy to a very confused octopus (Figure 1.4).

The leaky bucket!

Energy policy and generation are big policy issues, which would normally be considered beyond the remit of the client or the construction project team. However, I would say that any project for delivering sustainable buildings, whether new-build or refurbishment, should ask: where is the bulk of the power coming from? True attempts at sustainability should try to negotiate a supply contract to come from a renewable or as low carbon a source of energy as is available. It is important that when we talk about delivering a sustainable

Figure 1.5 The leaky energy bucket!

built environment the whole supply chain is taken into account. Ultimately, how efficient is the energy supplied to the building and what are its carbon implications from its source of generation and demand on natural resources?

Government and wider industry is waking up to this, but many still haven't grasped the concept, and others don't feel it can be done in time, but I am convinced that it's what we should concentrate on. If we're going to have a proper green deal, we need massive market intervention. Government and industry need to lead with energy conservation and efficiency as the priority. We need to plug the holes in the leaky energy bucket! (Figure 1.5)

The other important element is the significant impact that building controls can have on the energy hierarchy. (See Chapter 10 for information on reducing energy and getting control of it.) This is about making sure that everything is optimised and switched off at the right time, that equipment cuts out at the right temperature and operates within the right parameters. This is key to efficiency and to achieving steps towards the ultimate goal of sustainability.

References

Advertising Standards Authority (2002) ASA Non-broadcast Adjudication: Ecoflow Ltd. http://www.asa.org.uk/ASA-action/Adjudications/2002/4/Ecoflow-Ltd/CS_33701. aspx (accessed 13.8.2012)

Allen M, *Popular Mechanics* (2005) http://www.popularmechanics.com/cars/alternative-fuel/gas-mileage/1802932 (accessed 13.8.2012)

BERR, Department for Business, Enterprise & Regulatory Reform (2008) *Meeting the Energy Challenge: A White Paper on Nuclear Power*

The Consumer Protection from Unfair Trading Regulations (2008) http://www.legislation.gov.uk/ukdsi/2008/9780110811574/contents (accessed 13.8.2012)

Crabb, J (1997) *Field Test of Fuel Efficiency Magnets*; Exeter University Centre for Energy and the Environment

Department of Energy and Climate Change, DECC (2010) *The Green Deal; A summary of the Government's proposals*

Guardian Friday 17 February 2012, 'David Cameron in France to sign nuclear power deal'

Koplo D (2011) The Union of Concerned Scientists (UCS), 'Nuclear Power: Still Not Viable without Subsidies'

Malina, M. 'Bling Generation'. CIBSE Journal, January 2010. pp. 42–43. Available at: http://content.yudu.com/A1k3hn/CJJAN10/resources/42.htm

MORI 2010 Ipsos MORI/Cardiff University survey Public Perception of Climate Change and Energy Futures in Britain

Powell M. (1998) *Sceptical Inquirer*, Volume 22.1, January/February 1998 'Magnetic Water and Fuel Treatment: Myth, Magic, or Mainstream Science?' http://www.csicop.org/si/show/magnetic_water_and_fuel_treatment_myth_magic_or_mainstream_science/ (accessed 13.8.2012)

SDC Sustainable Development Commission (2006) *The Role of Nuclear Power in a Low Carbon Economy*

2 Planning ahead – the role of planning authorities

Local planning authorities can have a major influence on the development of sustainable communities and the built environment, both from a commercial and a housing point of view. But the planning system is often constrained by conflicting policy and resource priorities, as well as having to respond to, and take account of, the local community's views. Some would consider that planning in the UK can be over-democratic. The often-used phrase NIMBY (not in my back yard) describes a reaction from people who instinctively either want to resist change or perhaps oppose a major development such as a wind farm for either misconceived or legitimate reasons, depending on your viewpoint. A case of democracy in action? It is sometimes difficult to get the balance right. The problem is that we can only talk for so long about some of the very big choices that need to be made. Big issues and choices lie ahead, especially when it comes to major energy policy, the issues of natural resources, expanding population and their relation to the holistic view of sustainability. The scale of all of this is daunting when considered as one big project. However, this is unlikely to be considered as one grand plan; it is more of a collage which has grown over time, and will be dealt with as part of the evolving society we live in. Figure 2.1 reminds us of the scale when we look at a developed city.

A major influence?

The current planning system is highly decentralised, albeit guided by the recently revised National Planning Policy Framework. This is potentially good for local democracy, but it leads to a haphazard system where development is often speculative. Historically, there have been local and regional spatial plans, but at the time of writing the regional plans have recently been abandoned by central government. Therefore, except for large national interest projects which are led by central government and national infrastructure policy, all planning is done locally by professional planning officers employed by local councils. They are accountable to a community-elected body of councillors which comprise the local authorities planning or development control committee. Most planning decisions are determined

Delivering Sustainable Buildings: an industry insider's view, First Edition. Mike Malina.
© 2013 Mike Malina. Published 2013 by Blackwell Publishing Ltd.

Figure 2.1 Looking at the scale of the built environment makes planning very complex

under delegated powers by the officers, but if the plans are for larger developments, or potentially highlight a question of policy, then the planning committee will debate the proposals. The other important dimension is to show a public demonstration of the democratic process, where councillors can hear the views of the local community either by written representations or in person. This can be one of the most problematic aspects as there will always be two sides of the debate and it is often one of the only times that local people become involved in the planning process. My own experience was gained as an elected local government councillor in the 1990s, where I served on full planning committees and both unitary development and local plan subcommittees.

Decisions have to be made, and a lot is open to differing views of elected councillors, professional officers and the community they serve. There will certainly be different interpretations, even taking account of the professional guidance that is always provided. Local and regional forward planning is essential and it is unfortunate that government threw local authorities into turmoil with the scrapping of regional planning. Not to say that they were all good in the first place, but at least the bigger picture was being looked at.

Obviously, there are planning criteria identifying what kind of development is and isn't permitted, but local plans often do little more than ring-fence areas for different types of development – say, residential, industrial or retail. This created permitted development zones with specific policy guidance. Apart from that, the system tends to be ad hoc. It relies on developers coming forward with plans for development. For major developments, local planners will put forward a design brief, but they have no control over how things will develop overall in their local area. They might have a target, for example, of 800 homes by 2016, but executing this plan will depend on the vagaries of the market. Development can be, in many respects, little more than a random patchwork quilt.

In some senses, this is a good thing, because it allows each area to develop in its own individual way, but it is also highly inefficient from a resources and sustainability perspective. Infrastructure may well be developed after buildings are already in place, where clearly it should be developed before building construction, and this leads to public discontent. Significant new building projects create a need for roads, healthcare and education, for instance, which may not be fully developed and integrated when the plans are first drawn up. So any mechanism that can aid infrastructure and gain resources for the community should be made a priority, especially if this will also create benefit from a social and environmental dimension.

Planning gain

To help create infrastructure and more community resources, planners can draw on legislation such as Section (S106) of the Town and Country Planning Act 1990 (previously Section 52 of the Town and Country Planning Act 1971). This allows a local planning authority to enter into a legally binding agreement with a landowner for reaching an obligation associated with the granting of planning permission. This is referred to as planning gain, termed a Section 106 agreement. In Scotland it's a Section 75 planning agreement (Scotland Section 75). This creates a means of getting developers to contribute towards the community. If, for example, they build 30 houses, they must contribute so many pounds towards schools. It's a form of forward planning, and generates money from applications to fund infrastructure. The problem is that it is still ad hoc, and money from these agreements may not actually find its way to the local community. It's often put in a central pot that may get used anywhere.

Different governments have had different approaches for dealing with this question of planning. There are many arguments that can be used – for example, standardised and central planning would be more efficient, and create a focus for everything to happen in a coordinated way. From a sustainability point of view, it would also be more advantageous, since it would mean that councils, developers and utility companies wouldn't have to dig up the same piece of road six times to put in different services as more developments are created. It could be said that localism is good for fostering individual and local decisions. In my view, however, it is not good for sustainability as it doesn't utilise economies of scale and it's too ad hoc. It remains to be seen how the Localism Act, passed into law by parliament in November 2011 develops with regards the whole planning process.

As a case in point, when it comes to buildings such as new schools, they are all currently designed differently. That is justified as being good for the individuality of different communities. However, you could argue that it's a bad way of using resources. Surely, the needs of schools are roughly the same everywhere? Yet architects design radically different types of buildings for each locale. This means that there is no agreed national standard for design. It would seem to make more sense, from a sustainability and resourcing perspective, to standardise and agree a national design. This could form the basis for sustainable school buildings. They wouldn't all have to be exactly the same, as this really would take away individuality. However, it would seem reasonable to propose that the bulk of each school design should be similar, with some local individuality in terms of external appearance. Designing each separately seems to be a total waste. It's the same with hospitals and health centres. There is a lot of duplication of effort, and it slows the building process down. Some uniformity would allow for a lot more best practice to be used, and for a lot more off-site fabrication. This would speed the process up and deliver buildings in a more cost-effective manner.

Therefore, there should be a debate about the entire planning system. Is it working? These issues around duplication of effort need to be taken into account. There has to be a way to find a balance between the needs of each community, economic growth and the types of buildings we want in relation to their economics, their cost-effective delivery and the resource and environmental impact.

In the rest of this chapter, we will be looking at how planners could make buildings and communities more sustainable. But, no matter what is stated in planning documents the dilemma is still an issue as covered in Chapter 1.

The original and most widely recognised definition of sustainable development came from the Brundtland Commission report (UN 1987), *Our Common Future*, stating that 'Sustainable development is development that meets the needs of the present without compromising the ability of future generations to meet their own needs'. It is a shame that sustainability has become an overused word and has almost been lost from this original definition. Some may argue that this original description was too vague and did not provide a method for how sustainability could be made practicable. However, planners could have a very big role and influence in making a practical system work, although not the easiest of tasks. Whatever the development, whether urban or rural, nothing is truly sustainable and every development has an impact.

Figure 2.2 shows several acres of former agricultural land now to be covered in new housing. Could this be described as sustainable development? What are the impacts of a whole range of issues, for example water (see Chapter 7) and loss of more countryside.

Figure 2.2 Every development has an energy, resource and environmental impact

The disconnect between building control and planning

The building regulation systems (and the planning system, to a lesser degree), originate from the great fire of London in 1666. That catastrophic event, which destroyed 80% of the city, made the authorities and people realise that systems were needed so that precautions against fire spreading could be implemented. The first example of regulation was the London Building Act of 1667. This provided legislation on such vital issues as fire breaks. It also laid down that all houses were to be built of brick or stone. These regulations only applied to London, however, and provisions remained different in different parts of the UK right up until the 1984 Building Act of England and Wales. This was the first time that national legislation on building had been implemented.

A lot of people get confused about the relationship between planning and building regulation. They are in fact totally separate departments. Unfortunately, they often don't communicate with each other. Planning is all about the use of land and the creation of a wider infrastructure to serve a community and enable it to function. That's where it ends. Once planning permission is granted, the developer has to submit a building control/regulations application in order to start the construction of the building. The application contains all the details of the structural elements of the building, the thermal insulation, and the calculation of the building services to be used in relation to energy consumption. This application must now also include a building and commissioning programme for larger developments.

This system means that there is an important disconnect, because planning departments don't routinely actually check to make sure that what was laid down in the planning conditions is being fulfilled. Building control, on the other hand, doesn't check the planning side. The biggest problem is that, while in theory, buildings are designed and approved to be sustainable, it is often the case that these sustainable elements are lost in the construction process, when cuts and alterations to the budget mean that the finished building doesn't deliver what was intended. Many standards are not enforced. The bottom line is that many local authorities do not have the resources to actively enforce legislation.

The problem is that historically, communication between planning and building control doesn't happen effectively. There is also the practicality of enforcement – the two departments have separate enforcement officers. It has long been argued by the industry, and also by some local authority practitioners, that there should be generic enforcement officers who know both disciplines. This would make sense because of the resultant economies of scale, it would provide for better delivery of programmes, and would be less bureaucratic and more effective. So why isn't this happening? The difficulties are rooted in the past, and in different professional disciplines with their own body of knowledge and professional loyalties. In modern times, with more pressure on resources than ever before, this may be the solution to resourcing issues and to providing effective delivery of public services.

As well as land use planning, and hopefully planning for the infrastructure to serve these buildings, planners do have a role in creating sustainable buildings. There are two ways they can do this.

The first way is the design brief. This basically lays down the usual planning criteria such as appearance, size and density of a development, but it can also lay down factors such as the geographical orientation of a building. This can be quite influential in maximising passive solar gain and mitigating the effect of heat. Admittedly, this will work more effectively for an isolated building than for a development in a city, where there can be a heat island effect. Still, this is just one example of how sustainability issues could be addressed through the design brief.

The Merton rule

The second way of influencing sustainability is by setting planning gain for encouraging sustainable features and design. This stems from what is known as the Merton rule, which is named for the pioneering borough in which the approach was developed and adopted in 2003 (Merton 2003). Most local authorities have adopted it as the way forward, and it influenced national planning policy, being adopted in Planning Policy Guidance 22: Renewable

Energy which has now become Planning Policy Statement 22 (PPS22) and is now incorporated in National Planning Policy Framework.

The rule is that 10–20% of all new developments have to generate on-site energy via renewable or low carbon technologies. At first, this seemed like a good idea and was well received by the green lobby and by more enlightened building professionals. Developers were very resistant, because it would clearly cost them a lot more. Since then, however, it has basically become a tick box exercise. In practice, it became apparent that planners didn't know the difference between a good or a less desirable low carbon, renewable technology, or whether the technology was suitable for the application of the building and its use. There also seems to be a great deal of confusion as to what a renewable technology actually is. For example heat pumps are being labelled renewable as opposed to their true classification of being a lower carbon technology in application. More explanation and classification needs to be addressed as to what is truly sustainable, bearing in mind lifecycle and embodied energy costs. Too often, the planners weren't able to answer that question, and some debatable decisions have been made. For example, there was a trend in London and some other cities for biomass boilers. It was seen as a renewable technology, because it comes from a so-called renewable source – wood chips. But because these boilers were being located in a city, lorries were needed to transport the wood chips or wood pellets (which are worse because they are processed, creating more embodied energy), and the lorries clearly added substantially to the carbon footprint of the fuel. The buildings also need extra dimensions for storage of this fuel, and robust storage which is more resistance to fire hazards. Another question is, should we really be burning anything in a city, bearing in mind that it is meant to be a smokeless zone? Although the smoke content of the boilers is low, this is still unarguably 'burning'. Finally, planners were not making sure that what they were approving bore any relation to reality. In fact, many of the biomass boilers have never been fired. They might have been tested, but they were never fired. Conventional fossil fuel boilers, mainly gas, are being installed at the same time!

This is where the system overlaps with the energy hierarchy. The Merton rule tries to use this. The rule was a great idea, but it suffered from bad implementation. This is another example of the disconnect between planning, building control and building services specialists who could have a vital role in working with and educating all parties on developing the appropriate specification in the right circumstances. Also beneficial to building services professionals would be learning how planners and building control work. The more cross-sector and professional information exchange, the better.

We already have regulations to follow that will get more and more geared to energy saving in the coming decade. Photovoltaic panels are also used to satisfy the Merton Rule. This has led to grotesque situations. Because planners in the past didn't have technical building services knowledge, it costs a lot to satisfy the rules, but all it achieves is a gross distortion of carbon. In one real example, because there was no way of demonstrating energy saving in a proposed storage warehouse which didn't use much in the way of heat, planners decided to insist on photovoltaic panels. This created a £150,000 extra spend. That money could have been spent on carbon and sustainability savings if planners had better understood the opportunity to hypothecate, or offset, the energy emissions. It would be perfectly possible to spend the money on something totally different and offset the energy efficiently in a different way. The embodied energy in the solar panels ended up contributing little, whereas the money could have gone to landscaping or something else for the public good. It could have benefitted the public via Section 106, or could have been put into the local communities carbon reduction by way of contribution to any appropriate community project.

These policies have also, I believe, been distorted for setting a minimum level for on-site renewable energy generation. Some planners have perhaps decided that they wanted to promote a policy for encouraging the take-up of renewable energy generation, almost looking to influence the take-up and create more of a local market for it, thus moving away from the original intention of creating a more sustainable system of working through the energy hierarchy, which should be the overriding aim for maximum impact on carbon reduction.

Training for planners and building control officers

In short, it is good to have a requirement for renewable energy as part of the planning process, but it has to be implemented more effectively. If we are going to introduce new technologies, we have to understand their application. How should that happen in practice? To begin with, planners need to be educated and trained, or at least to work more closely with consultants and specialists to achieve the best possible outcome for a project. In the planning process, the emphasis should be on the energy hierarchy, to minimise energy use. In fact, the Merton Rule can actually create more energy use. Unfortunately, under the current system, the planners may argue that this was nothing to do with them, but a matter for building control departments. This has to be another argument for more streamlined working between the two constituents.

A further flaw in the system is that planners haven't looked at building management systems (see Chapter 10 for more information) to date, which means that these have been left for building control to deal with. Building control are only looking for minimum compliance with legislation, and this, in my experience, is often missed or overlooked. If planners were involved in this aspect of design, they could stipulate that they wanted wider environmental and building management control considerations in the design brief, as a requirement for meeting carbon reduction aims and objectives. This would be one of the most effective measures to get, in effect, two means of compliance covering the planning and building regulations angles.

Climate change skills training for planners

(Sustainability East, 2012)

A great start has been made in the East of England and a number of other regions. In the East, a joint project funded by the government Climate Change Skills Fund has led to a series of events in which Sustainability East sponsored 'Climate Change – Training for Planners' working with the Town and Country Planning Association (TCPA). This ran as several sessions and was well attended by local authority officers, including senior planners, sustainability coordinators, and a good number of elected councillors, who often make important planning decisions on their local authority planning committees.

The sessions topics covered included:

- an introduction to planning and climate change
- evidence base and policy development
- delivery and planning application determination
- climate change mitigation solutions for buildings and developments
- climate change risk and adaptation
- stand-alone technologies for mitigation and adaptation
- study visits to an eco-community development in the East of England
- training the trainers.

Of particular interest was the session on delivery and planning application determination, which covered key elements of project development and an introduction to delivery. It also encompassed the determining of planning applications and building regulation applications in relation to climate change policies, particularly the interpreting of climate change and energy statements. The importance of monitoring actual approvals was discussed, including such aspects as what elements need to be monitored; measuring the right things; and using datasets and support tools. It also covered negotiating with developers on climate change issues, including Section 106 agreements, as well as the role of pre-application discussions in improving outcomes on climate change issues. Finally, the approach and structure of the elements of stakeholder engagement on developments was explored.

These sessions involved practical exercises, which should be welcomed in equipping planners with the experience to negotiate stronger requirements for planning gain encouraging low carbon developments.

One disappointment in the process was the lack of attendance of building control officers and practitioners. This, I believe, should be the next stage to bridge the still real disconnect between planning and building control departments and to help the process for joined-up mechanisms to ensure delivery and verification of low carbon developments for the future. It would also be useful for building services practitioners to be brought in to give some practical feedback on the realities of the true performance of low carbon technologies (LCTs). Different designs and developments will favour more appropriate applications of renewable or LCTs. The important point here is that, so far, the planners and building control officers have not had the experience or knowledge to make a choice of which technologies could be deployed. They are unlikely to ever get this in any real depth. This is why CIBSE should be making more effort to engage with the TCPA and RICS.

In the East of England we are starting to do just that. Joint events and meetings with RIBA and RICS are taking place on themes of mutual interest. Sustainability in the built environment is the theme that makes this possible, because the holistic nature of the issues require joined-up thinking. This is also happening at Sustainable Built Environment East (SBEE), which is covered in more detail in Chapter 5.

Planning for the future

Rather than waiting for planners or for building control to respond, once a design is drawn up, it makes sense to seek the opinion of the planners from day one. It mitigates future problems, is cost effective and gives developers the chance to explain to the planners their philosophy and approach. If there was a danger that they were going to apply the Merton rule without understanding the holistic design intention, designers could explain their thinking from the start and not find themselves the unwilling participants in a tick box exercise. This should make it possible to show more effective solutions than the ideas that might otherwise be imposed. Such early collaboration should be the first action in creating a far more positive process. At the same time, making links between building control and planning would be advantageous. The other thing designers should do is to collect as many examples of similar projects as possible to build a case and to show that their proposed solution works and delivers the carbon savings required.

Localism is a good thing in principle for local democracy and community involvement, but we must never lose sight of the big picture. Strategic regional and national planning is vitally important and must be joined up.

New buildings and infrastructure are important, but the sharing of resources and existing infrastructure will be one of the main challenges for the future. The vast majority of this will need significant planning and reengineering to take priority along with the massive task of refurbishment and upgrading of existing buildings (Figure 2.3). Key examples will be the linking of energy producers, whether large power stations or smaller producers, and linking processes with waste heat, district heating and waste to energy developments to appropriate applications to serve businesses and the residential community.

Figure 2.4 shows a good example at British Sugar's Wissington factory in Norfolk. A combined heat and power (CHP) plant, produces steam and electricity for processing and also exports over 50 MW to the local electricity grid, which can serve a population of 120,000 people. A considerable amount of the flue gas which would otherwise go up the chimney is diverted to the

Figure 2.3 Retrofit and refurbishment will be key in making an impact

Figure 2.4 Industrial combined heat and power production shared with the community

nearby greenhouses that form the largest UK classic tomato growing site, covering 27 acres, providing heating and CO_2 which are used to promote plant growth.

The other important task that needs to develop in a wider context is for the building services' professional bodies to do more to engage with the planners' professional bodies. They should also be looking to influence both local authority officers and elected councillors to establish a good rapport and to help them to understand the important issues that affect the built environment. Holding joint seminars and conferences could bring this closer to fruition and lead to a better way of working. We would all gain a greater understanding of all our professional roles in working towards delivering a sustainable built environment and a lower carbon future.

References

(Merton 2003) The Merton rule http://www.merton.gov.uk/environment/planning/ planningpolicy/mertonrule/what_is_the_merton_rule.htm (accessed 14.8 2012)

(PPS22) Planning Policy Guidance http://www.communities.gov.uk/planningand building/planningsystem/planningpolicy/planningpolicystatements/pps22/ (accessed 14.8 2012)

(Sustainability East, 2012) http://www.sustainabilityeast.org.uk/index.php?option=com_ content&view=article&id=76&Itemid=77 (accessed 14.8 2012)

Town and Country Planning Act 1990 http://www.legislation.gov.uk/ukpga/1990/8/ section/106 (accessed 14.8 2012)

Town and Country Planning (Scotland) Act 1997: Section 75 Agreements http://www. scotland.gov.uk/Publications/2010/01/27103054/2 (accessed 14.8 2012)

(UN 1987) Report of the World Commission on Environment and Development: Our Common Future http://www.un-documents.net/wced-ocf.htm (accessed 14.8 2012)

3 Legislative overview and meeting your legal obligations

Global targets and local actions

Legislation on carbon reduction, mitigation of climate change and use of low carbon and renewable energy sources is having a huge influence on today's construction industry. This, coupled with specific legislation, from building regulations to voluntary environmental standards, is beginning to change the environmental and energy performance of the built environment.

The legislation implemented here in Britain has its origin in worldwide agreements, and in some cases in international treaties which were signed well over a decade ago. High level agreement by several national governments on the need to mitigate the effects of climate change was first reached as far back as 1988. It was in this year that governments established the Intergovernmental Panel on Climate Change (IPCC).

The IPCC was the result of a growing consensus among scientists that global warming would have devastating effects on the planet and its population. The first report from the IPCC was published in 1990, and its main recommendation was that governments around the world should agree to work together to start combating the causes of global warming.

The UN Framework Convention on Climate Change (UNFCC) was signed at the Rio Earth Summit in 1992. The Convention established an annual Conference of Parties (COP) as the mechanism for implementing actions, and the first goal it established was that all signatory countries should return their emissions (carbon and other greenhouse gases) to 1990 levels. Since that initial goal, the Convention set up the Kyoto Protocol as the mechanism to deliver further global carbon and greenhouse gas reductions.

Under the Kyoto Protocol, the European Union (consisting initially of 15 nations) was given a collective target of reducing its greenhouse gas emissions to 8% below 1990 levels by 2012. The new states that joined the EU were also included in this target on their accession to the EU in May 2004. A 'Burden-Sharing Agreement' allocates this target between the member states, and the UK was given a target of a 12.5% reduction.

The UK government has gone beyond this requirement, with a number of its own ambitious targets. Initially, the government produced an Energy White Paper (DTI 2007) which set out the objective of cutting UK carbon

Delivering Sustainable Buildings: an industry insider's view, First Edition. Mike Malina.
© 2013 Mike Malina. Published 2013 by Blackwell Publishing Ltd.

emissions by 60% against 1990 figures by 2050. However, this was replaced by a higher target of an 80% reduction by 2050 and introduced in November 2008 the Climate Change Act, giving the government a legally binding framework for ensuring that it meets its commitments to mitigate the effects of climate change. The UK is the first country in the world to introduce emissions reduction targets as law. This is highly ambitious and so far looks a long way off with regards to actually achieving anywhere near this objective.

The background to raising this target came from the Committee on Climate Change (CCC), which was set up as an independent body to advise the government on setting and meeting carbon budgets and to advise on ways of preparing for the potential impacts of climate change. In October 2008 in a letter (Ecchinswell 2008) to the government, the CCC advised the government to raise the CO_2 reduction target from 60% to 80% after careful analysis of scientific understanding of the impacts of global warming indicated greater and faster action would be required.

The government believes that the 2050 target is achievable through increased use of renewable energy sources in the long term, though energy efficiency will be the main method of cutting the country's emissions in the shorter term. The CCC has identified a number of areas where carbon savings can be made, and recognises energy reduction in buildings as playing an important part.

The European influence

In Britain, we have felt the impact of international targets on carbon reduction through a number of legislative measures such as the European Energy Performance of Buildings Directive (EPBD). Studies for the European Commission show that 40% of Europe's energy is used on heating, lighting, cooling and running homes and offices. Research also suggests that if all potential energy savings were achieved, the EU could be consuming 11% less final energy than it does today, and saving 4% to 5% of its current carbon emissions. Hence, the EPBD focuses on improving the design of buildings for better energy efficiency, while setting minimum standards of energy use and long-term energy performance (EC 2003).

The current version of the EPBD was introduced in 2005 and has since been updated to form the recast EPBD to be introduced for general implementation from 9 July 2012. This is being introduced to clarify certain aspects of the earlier Directive, and, extend its scope, strengthen certain provisions, and give the public sector a leading role in promoting energy efficiency.

It is primarily designed to reduce the energy used in commercial and domestic buildings, with a four-step approach:

(1)   Establish a framework for calculating the energy performance of buildings. Methods used for this calculation differ between member states. In the UK, we use the Standard Building Energy Model (SBEM) which is based on modelling software for buildings.

(2) Set minimum standards of energy use and performance for new and existing buildings. These standards have to be demonstrated at the design stage.

(3) Introduce energy certification of buildings. These are known as Energy Performance Certificates (EPCs). Accredited assessors examine a building's energy use, issuing a graded certificate (A to G) and including advice on how to improve energy performance.

(4) Introduce regular checks on air conditioning and heating installations.

The Building Regulations and standards – England and Wales

The Building Regulations themselves set out the legal requirements and procedures for compliance, although they actually give very little practical guidance as to how you should meet your obligations technically.

For this, the government publishes a series of documents, known as 'parts' for technical guidance, and these are labelled as approved documents. These form parts A to P, each as a separate approved document containing guidance for some of the most common areas of work, setting out how to formally achieve compliance with the regulations. The approved documents themselves are not legal requirements and they state that it is possible to comply with the regulations using other sources or means of demonstration. This could be gained from industry standards or manufacturers' guidelines. However, accepting the methods and solutions from the approved documents would normally result in compliance with the Building Regulations themselves.

Part A – Structure
Part B – Fire safety
Part C – Site preparation and resistance to contaminants and moisture
Part D – Toxic substances
Part E – Resistance to the passage of sound
Part F – Ventilation
Part G – Sanitation, hot water safety and water efficiency
Part H – Drainage and waste disposal
Part J – Combustion appliances and fuel storage systems
Part K – Protection from falling, collision and impact
Part L – Conservation of fuel and power (four documents)
Part M – Access to and use of buildings
Part N – Glazing – safety in relation to impact, opening and cleaning
Part P – Electrical safety – dwellings
Approved document for Regulation 7 – Material and workmanship

All these documents can be downloaded from: http://www.planningportal.gov.uk/buildingregulations/approveddocuments/
These are valid in England and Wales only.

Scotland and Northern Ireland

Scotland and Northern Ireland have a similar system but a different set of documents with particular points developed to meet devolved legislation and standards.

In **Scotland** they are referred to as Building Standards, and guidance is provided by technical handbooks, as non-domestic handbooks and domestic handbooks, consisting of the following:

Section 0 – General
Section 1 – Structure
Section 2 – Fire
Section 3 – Environment
Section 4 – Safety
Section 5 – Noise
Section 6 – Energy
Section 7 – Sustainability
Appendix A – Defined terms
Appendix B – List of all standards and publications referred to within technical handbooks
Appendix C – Index
Supporting guidance

All these documents for Scotland can be downloaded from:
http://www.scotland.gov.uk/Topics/Built-Environment/Building/Building-standards

In Northern Ireland the regulations and technical booklets consist of:

Part A – Interpretation and general
Part B – Materials and workmanship
Part C – Preparation of site and resistance to moisture
Part D – Structure
Part E – Fire safety
Part F – Conservation of fuel and power
Part G – Sound insulation in dwellings
Part H – Stairs, ramps, guarding and protection from impact
Part J – Solid waste in buildings
Part K – Ventilation
Part L – Heat-producing appliances and liquefied petroleum gas
Part N – Drainage
Part P – Sanitary appliances and unvented hot water storage
Part R – Access and facilities for disabled people
Part V – Glazing

All these documents for Northern Ireland can be downloaded from:
http://www.buildingcontrol-ni.com/sections/?secid=5

Applying the EPBD in the UK

Throughout the nations of the United Kingdom, the EPBD is interpreted into the specific nation's documentation.

I will concentrate on England and Wales, but much of the application of the EPBD will be common to the rest of the UK as well.

In England and Wales, the original EPBD was introduced into national law through a new version of Part L of the Building Regulations. On 6 April 2006 new versions of two key elements of the Building Regulations for England and Wales – Part F (Ventilation) and Part L (Conservation of fuel and power) – came into force. This has been continued with the 2010 revisions. The revisions encompassed in these new regulations included measures to reduce the energy consumption of new and refurbished buildings, while also ensuring adequate ventilation. Parts L and F of the Building Regulations are closely linked, reflecting the importance of adequate ventilation in tightly sealed buildings, whilst emphasis is placed on proving actual rather than predicted performance.

Part L 2006 and 2010 is a practical approach to delivering the four key elements of the EPBD and includes a number of important new requirements for commercial and domestic buildings:

■ targets for carbon dioxide emissions resulting from energy use in buildings
■ a standard method of calculating building energy performance
■ energy performance certificates and regular checks on heating and cooling systems

From 2010 it was announced that a three-yearly review cycle will be carried out aimed at progressively reducing emissions. This is covered later in this chapter.

The 2010 legislation set targets for reducing CO_2 emissions against 'similar' buildings constructed to Part L 2006 standards and a 25% reduction in carbon dioxide emissions compared to a similar building constructed to Part L 2006 standards, which is supposed to deliver a 40% reduction relative to 1995 standards. At the time of writing the government has just released the consultation for the 2013 revision to the Building Regulations.

In order to target different types of properties Part L is broken down into four sections, or approved documents (ADs). These are as follows:

■ Approved Document L1A (ADL1A) – new dwellings
■ Approved Document L1B (ADL1B) – existing dwellings
■ Approved Document L2A (ADL2A) – new non-dwellings
■ Approved Document L2B (ADL2B) – existing non-dwellings

Technical documents in support of Part L were also revised in 2010 and comprise:

- Non Domestic Building Services Compliance Guide 2010
- Domestic Ventilation Compliance Guide 2010 and Domestic Building Compliance Guide 2010.

Building performance

Energy performance for dwellings and non-dwellings is calculated in terms of the building's overall energy use, expressed in terms of carbon dioxide emissions. Designers must establish a target CO_2 emission rate (TER) and then calculate the projected emission rate for the actual building, known as the building CO_2 emission rate (BER). The BER must not exceed the TER.

The TER is found by calculating the emissions from a notional building that complies with the minimum Part L regulations, and then making reductions in energy consumption.

In calculating both the BER and the TER, designers must use the National Calculation Method, which is approved by the EU. For dwellings, this is the Standard Assessment Procedure (SAP) and for non-dwellings this is the newly introduced Simplified Building Energy Model (SBEM). A number of software packages based on SBEM are now available on the market.

Existing buildings

While most of the impact of the new Part L will come from new buildings, it does also apply to existing buildings under certain circumstances.

Consequential improvements

This applies on buildings with a floor area of over $1000\,m^2$ where there is a new extension or an increase in the capacity of a fixed building service. Under these circumstances, the principal works have to comply with the Part L guidance, and improvements to the rest of the building as a consequence of this new work will also have to follow the guidance. This includes bringing thermal elements (floors, walls and roofs) up to 2010 standards, and any services that are over 15 years old will have to be upgraded.

During the 2010 consultation and hopefully during the 2013 implementation the consequential improvement rule will be extended to domestic properties as well as widening the scope of commercial buildings by lowering the $1000\,m^2$ threshold to cover all buildings. I believe that this is vital if we are to come anywhere near to meeting the targets for energy efficiency and carbon reduction. As most of the buildings have already been built, the focus must be on existing buildings and how these are refurbished and retrofitted to achieve anywhere near the carbon reductions that are required.

Extensions

Any extension that is over 100 m² *and* greater than 25% of the existing building counts as a new building and will have to achieve a satisfactory BER. Smaller extensions, down to 30 m², have to make 'reasonable provision' to meet the standards through measures such as using controlled services and thermal elements that meet the standards laid down in ADL2B.

Material change of use

ADL2B gives a number of examples of material change of use. These include conversion of a non-dwelling to a dwelling; conversion of a private commercial building to a public building; and addition of rooms. Detailed information is provided in Section 4 of ADL2B.

Material alteration

This applies to any work that would lead to non-compliance of a building or service, which previously did comply. Or, with a building or service that did not comply, where any proposed changes would worsen the non-compliance.

Work on a controlled service or fitting

Controlled services and fittings are defined as services or fittings covered by Parts G (Hygiene), H (Drainage and waste disposal), J (Combustion appliances and fuel storage systems), L (Conservation of fuel and power) or P (Electrical safety). These include windows, roof windows, roof lights, entrance doors, vehicle access doors and roof ventilators.

Continuous change – Part L 2010–2013–2016 and a new EPBD

In October 2010 a new set of revised building regulations came into force. This started a three-year periodical review process that will see progressive changes in 2013 and 2016. One of the reasons for this update is that the UK government has set higher targets for carbon dioxide emission reduction than other countries and wants the requirements of Part L to reflect this.

The government also wants to improve compliance with this legislation, which has been lacking largely due to a lack of knowledge and personnel in the building control profession. Furthermore, updates have been made to the SBEM to allow for advances in technology and materials.

The points below are a summary of the main changes to Part L from 2010.

- A 25% aggregate reduction in CO_2 emissions is required for all buildings compared with 2006 levels. This means that some buildings will have to meet less than 25% reduction, others more, depending on building type.
- Recognised software will be used to comply and list the main features that will enable a building to meet its energy performance targets.

- All design stage submissions to building control have to be accompanied by a specification. Designers also have to submit a commissioning plan at the outset of a project.
- There is a revised limit on passive solar gains for non-domestic buildings.
- A fuel-based TER has been introduced to help to improve energy efficiency.
- Emissions from electric heating systems are capped at the same rate as for oil heating.
- The SAP calculations accuracy will be improved by moving from an annual to a monthly calculation system with the inclusion of updated weather data.

The Green Deal (also covered later in Chapters 4 and 13) will probably form a significant part of any revision to Part L in 2013 and at the time of writing a consultation exercise is under way. The 2013 regulations consultation propose a further reduction of 8% for emissions in new dwellings and 20% from new non-domestic buildings. An updated version of SAP and SBEM will be made available to test the impacts of these new measures.

Taking the energy performance of buildings to the next level

In January 2007, the EU Commission introduced a climate and energy policy which includes targets known as '20-20-20%': the reduction of energy consumption and greenhouse gas emissions and increased share of renewables by 2020. The built environment is viewed as a key source for its targeted carbon savings. The EPBD is the main pan-European tool for delivering energy savings in the built environment. Its main aim is to encourage cost-effective improvement in the overall energy performance of buildings.

A new and recast *EPBD 2010* repealed and replaced the *EPBD 2002* from 1 February 2012 and must be implemented into national legislation by 2013. This will see more buildings covered by the legislation, higher efficiency targets and redefinitions of energy efficient technologies.

This recast is very different from simply amending the existing Directive. The recast will produce an entirely new legal document, which will replace the existing Directive. The EU Commission has decided on this course of action 'to ensure clarification and simplification of certain definitions and provisions of the current Directive'.

The Commission proposes a range of changes, including the following:

- The $1000\,m^2$ threshold for new and existing buildings (Articles 6 and 7) will be removed. This means that all buildings will be covered by the Directive regardless of size. The current $1000\,m^2$ threshold excludes 72% of Europe's building stock.

- The definition of heat pumps will be amended to include all types: water, geothermal and air source (Article 2/14).
- Inspections of just boilers will be replaced by inspections of whole heating systems. All heating systems with a boiler of an effective rated output of more than 20 kW should be inspected (Articles 13, 14, 15).
- An updated benchmarking tool for calculating cost-optimal standards will be revised to build on the standards set up (Articles 3 and 4) in the first EPBD.

Member states will have to impose penalties for infringement of national provisions of the EPBD (i.e. infringement of whatever law in each country has been used to introduce the provisions of the new Directive; in the case of the UK this would probably be a new Part L of the Building Regulations).

Legislation levels, change and enforcement – opinion

One of the questions that is often asked about sustainability is: why do we need so much legislation to drive it? It sometimes seems that government is determined to use more stick than carrot when it comes to achieving lower carbon emissions from our built environment. However, there is also the option for businesses to view legislation as 'carrot', in so much as it gives them an opportunity to demonstrate their commitment to their principles and to the good of the nation and society.

Burdensome as it can seem at times, in fact, legislation such as building regulations has to meet strict criteria on proportionality. Legislation has to go through a regulatory impact assessment, and any proposed changes must be shown to be reasonable in their effect on business. It is true that there are sometimes unforeseen consequences from legislation, but government can then revise it to make it more equitable.

The critical issue with building standards is: who takes on the risk for the whole building? Historically, there has been a fragmented approach in construction, but working out the issue of liability has to be addressed to make the industry more sustainable. There needs to be an integrated team, shared responsibility and liability. Then, we'll start to see improved procurement processes, and better ways of working together, guided by the need for clients to have more sustainable buildings. It's about the whole lifecycle. At the moment, the people who build the foundations wouldn't, by and large, even think of worrying about their client's heating costs. They just wouldn't see it as their area. But we will have to move towards a more holistic approach, and we are starting to get there through a combination of legislation and increasing standards. A lot of this legislation is coming from the EU. The way in which we know how energy performance legislation moves building users into performance monitoring and outputs will be the critical element. The team who build it are going to be much more responsible for output over time.

In Scotland, there is a different system of building control, governed by the Building Act. This requires parties who wish to design or construct a building to complete a system of notification through a building warrant scheme. This splits activities into two parts, so that the design and construction can be warranted separately. What tends to happen is that the architect (or other responsible person) will go round and collect all the individual certificates from all the key parties before the building itself is warranted. This means, effectively, that the whole building is warranted. This creates a kind of internal enforcement mechanism that seems to work: the architect, or whoever else warrants the building, becomes the policeman of the system. The system is therefore patrolled at no cost to the public purse.

Against that background, the governments will keep moving in that direction because the only way to reduce carbon emissions is through the improved quality of buildings.

There is indeed a lot of legislation, much of it about standards and performance. The legislation is largely coming from Europe. So how is the industry coping?

The fact that legislation seems to be changing so much is a cause for concern within the construction sector. In 2010, for instance, we saw a new version of Part L, as well as a re-launch of the EPBD. Changes have also been made to the government departments responsible for building regulations and sustainability (from the Office of the Deputy Prime Minister of the last government to DEFRA, the Department for Food and Rural Affairs, to the most recent, DECC, the Department for Energy and Climate Change) has also added to the confusion.

The response of the industry can basically be divided into two elements. One is that too much legislation gets ignored. That is certainly happening and is evident by the lack of enforcement. The second element is a curious problem, not necessarily unique to construction. Legislators are concerned about legislation, getting it done and on the statute books, but often fail to see if it is successful once it is enshrined in law. Similarly the EU's emphasis is concentrated particularly around resource and energy efficiency, but the drive is to get the laws passed rather than to see if they are actually working.

The key to these issues is enforcement and feedback. Firstly we need to ensure that the legislation is being implemented and then there need to be proper feedback loops from industry, so we can understand the long-term effects of the laws and their impact. We need a lot of feedback on whether or not things are working, and a commitment from governments to address or adjust the way legislation operates. That's not there at the moment and we are suffering from a clear lack of joined-up government thinking.

Joined-up government?

Government is only joined up to a degree. We think government is a single body, but it's not. It's full of different departments, all creating their own legislation, and to some degree protecting their historic territory. In terms of the construction sector, having a chief construction advisor who is tuned into the issues is vital. We're now starting to see a Green Construction Board developed, as part of the ongoing work of the chief construction advisor, which will look at whether industry can deliver on the sustainability agenda. Some of that work will feed into legislation, tying together the strands from across government to ensure that they are coordinated and working.

The lack of enforcement can also be understood in terms of the sheer amount of legislation. Quite simply, there's just too much going on at the moment. There is an argument for a quieter period for all this change to bed down, but that has to be balanced against the fact that the industry really needs to change still further. Do you stand still and let it be absorbed or move forward and see what happens? It's a difficult question and something that needs to be addressed.

At the moment, the industry is quite literally being battered in a whole load of areas. The danger is that, under the weight of the unachievable, the temptation will be to take no notice at all. There was a big drive on legislation during the last part of the Labour administration that ended in 2010, but we now have a totally different government, with a very different approach to spending and potentially to legislative intervention. This has got to have an impact on enforcement. That said, there certainly appears to be a significant degree of consistency between the two governments with regards to their sustainability policies, which are broadly the same. Obviously, especially considering that the Labour government was superseded by a coalition of two different parties, there would have been policy issues to resolve internally, but overall there has been a remarkable degree of consistency in recent years. The current government does seem to want to roll back some of the legislation and possibly enforcement; this may also be because of there are fewer resources to police it, which can largely be ascribed to the pressing financial crisis. There is no mainstream political party in the UK that refuses to believe that reduction of carbon emissions is important. That won't change, and the issue will only move up the policy agenda over time. Sustainability, the security of supply of energy and the various associated consequences are now in the national and international psyche. Politicians know that the business community, by and large, won't make decisions on the basis of concern about the melting icecaps and the fate of polar bears. However, they also know that the sustainability agenda has produced a critical opportunity for businesses to lead on the development of improved performance through productivity. That makes sustainability interesting to even the least sentimental and may well be a major future focus, emerging as the real opportunity for change. It's certainly an opportunity for the construction sector, which has long been working to become a lot more efficient.

Looking forward

The only guarantee, as far as regulation on energy efficient and sustainable buildings goes, is that there will be change on a regular basis. The government even now includes a 'Future Thinking' section in proposed legislation to indicate where the next stages might take us. Currently, there is an overall goal of an 80% reduction in CO_2 by 2050, but it will involve a step-by-step process to get there. The government has also seen the importance of a single voice on sustainability issues, and the development of the DECC, which links energy with climate change, seems to be a step in the right direction for achieving some joined-up government.

The 80% figure is an enormous challenge, but it is achievable. People worry about the cost or the type of technologies that we'll have to use. However, 2050 is still some way off, and the government has put in milestones along the way to that goal, such as a 26% target for 2020. People who are sceptical now are going to see growing evidence of the negative effect and impacts that this country will face because of the effects of climate change, and that will certainly concentrate minds around the need for change. Also, by 2050, we may well be using many technologies that haven't even been invented yet. This, in part, is why today's legislation is so important – to act as a driver for innovation in the sustainable buildings sector and to encourage seed funding for research and development of new technologies and techniques.

Certainly there is more legislation in the pipeline, particularly around energy and resource efficiency – these are hot topics in Europe. Things are going to move forward at the same fast pace, at least for the foreseeable future. The only thing that will slow it down will be other priorities, but there don't seem to be too many of those currently. A lot of what the current government is doing doesn't take up much parliamentary time, so there is time for it to go through and become law. The same is true in Europe – it is dominated by the financial crisis, but that's only one element. There are still lots of directorates beavering away at legislation, driven by Europe's collective signing of the Kyoto treaty to deal with the issues of climate change.

The outlook is good – we do currently have a lot of problems in companies financially, but they are outside our control. We can't manage the crisis. But there are underlying long-term issues, at the centre of the carbon reduction agenda, that businesses can have an impact on. The heating and cooling of the built environment is an enormous part of the equation here. People won't want a reduction in lifestyles; we've got to improve them if anything. They will look to the industry to do these things in a more energy- and operationally efficient way. These issues are key to the whole process now and in the future.

Enforcement

Enforcement of Part L of the Building Regulations has been an issue since the regulation's introduction in 2006. There are sections of the construction industry that believe that since enforcement of Part L for new buildings has been lax (particularly when compared to regulations on fire safety and disabilities access, for example), the rules on energy use and sustainable buildings can safely be ignored. This situation needs to change, and the government claims to be determined to increase enforcement.

There is certainly an emphasis in government now on getting the building controls bodies more on board with enforcement of Part L. The training of building control officers to date has been largely around safety, structural issues and drainage. Energy is different, because you can't see it. Education will play an important part in increasing enforcement, and training of building control officers is vitally important to make this happen.

A number of campaigns have tried to 'show' energy, for example the Carbon Trust's advertisements showing CO_2 coming directly from a car or a building; other pictures showing government office buildings with the lights left on all night have been used to 'name and shame'. Energy performance certificates (EPC) and display energy certificates (DEC) will also show the realities of energy use. When you get the data that is shown on a DEC, it makes very transparent how much energy a building is using. Of course, a big energy user isn't necessarily a poorly designed or operated building; you have to understand what's going on inside that building. It could be argued that DECs should be widened from use only in the public sector so that more organisations show exactly how much energy they are using.

Tougher enforcement may be needed to get the message across that government is serious about Part L and its targets for reducing energy use in buildings. When a few of these cases to go to court and fines are imposed, people will realise that they need to adhere to the rules. The courts need to set a precedent. Reducing energy waste and carbon emissions are issues that affect our health, the economy and the wealth of the country too.

An example of the industry being proactive to raise standards comes from the B&ESA (formally the HVCA) which formed The Building Engineering Services Competence Accreditation Ltd (BESCA) to undertake competent person assessments under Part L, creating an assessment body to do just that. This means that companies can self-certify their installations to comply with the building regulations and go through a simple process of notification to the relevant local authority where the work has taken place. At the time of writing, there has been a comparatively low take-up, which is directly attributable to the low levels of enforcement of the relevant legislation. As things stand, the authorities will only enforce the law if there are obvious breaches in the sign-off of the work. There is a big caveat to this, however, which is that there will only be an investigation if the breach is reported, or something goes wrong that makes a knock-on effect obvious in a building's operation. In effect, the current system relies on whistle-blowers, and there aren't too many of those in the industry. This creates a farcical situation where the minority of companies who invest resources in doing things correctly find themselves uncompetitive and disadvantaged as a result.

Some final thoughts

In spite of the importance of legislation, it can be argued that legal enforcement is the worst reason for ensuring that buildings are energy-efficient, particularly since regulations tend to set relatively low benchmarks. If people only do things because it's the law, I would say that's a very short-sighted approach. It makes economic sense to design and construct sustainable buildings because we can't imagine that we will have cheap energy forever. We can't keep thinking that we will always have pipes of oil and gas coming into this country to keep our appetite for energy fed. We must live within our means. That is the harsh reality, and the government is surely aware that security of supply is a growing area of concern along with the effects of climate change.

One ridiculous situation currently exists in Regulation 47 of the Building Regulations. This relates to certification of compliance in areas such as commissioning, building pressure testing, carbon emissions calculations and the issuing of self-certification scheme compliance and energy performance certificates. These are legal requirements and, technically, not following these procedures is a contravention of the regulations. However, Regulation 47 states that contravention is not an offence. So it is unlikely that a local authority building control officer will actively prosecute for non-compliance. Hence, since 2006 I know of no prosecution in this area. This is why a culture of non-compliance has evolved. So the industry is largely ignoring many of these requirements. This is scandalous, and all governments in recent years should be ashamed of this mess; so many opportunities have been missed in achieving proper standards and the achievements for carbon and energy reduction will have been missed.

Can governments be consistent in the future, developing the legalisation and most importantly making sure that enforcement takes place? One idea to lessen the resource load on central government is for the industry to take responsibility for the future development of the technical elements, for compliance with the building regulations, as detailed in the approved documents. This could follow the same model that's been used in the electrical sector for many years, in taking charge of the development of the wiring regulations.

We need good case studies to demonstrate that sustainable construction shouldn't be done simply because there are legal obligations; that is just the starting point. At the moment, it is sometimes difficult to prove energy use in a building once it is occupied and operating, but as we bring on board new technologies and testing mechanisms that are less intrusive we can gather more evidence. My hope is that we always have innovators who can lead the way by example.

People often talk about payback of sustainability, but there are many things we do that don't 'pay back', and you simply can't measure sustainability exclusively in this financial way. This is a difficult issue, because it is true that at the moment some of the aspects of sustainable design cost us more money in the short term. However, the long-term payback is that it will put us in a healthier state environmentally and economically. At the moment, we don't always see the true costs of our action, or our inaction. If we had proper carbon accounting, we could demonstrate the real cost, to our economy and our health, of pursuing a non-sustainable agenda.

References

Ecchinswell, Lord Turner of, (2008) Letter to the government. Available at: http://downloads.theccc.org.uk/Interim%20report%20letter%20to%20DECC%20SofS.pdf (accessed 14.8.2012)

Energy White Paper: Meeting the Energy Challenge, (DTI) Department of Trade and Industry, published 23 May 2007

European Commission, (EC 2003) Directorate-General for Energy and Transport, B-1049 Brussels http://europa.eu.int/comm/dgs/energy_transport/index_en.html (accessed 14.8.2012)

4 Paying for it – the finance question

The current financial system is geared to short-termism. To demonstrate this, we only need to look at the volatility of the stock markets. It seems to be human nature to measure only up to the horizon. In business, we seem to concentrate on looking at annual budgets rather than over the long term. This short-termism has created, for example, the pensions time-bomb and a host of other problems, because people haven't looked over the longer term and invested for their future.

When addressing the relationship between finance and sustainability, we could start with those professionals working in the financial sector. I believe that they are largely unaware of the rationale and pressures for sustainable development and its relevance to their work. Equally, those of us who are engineers and sustainability practitioners seem to commonly overlook how vital the financial sector is for our work and progress.

Short-termism – damages sustainability

Many companies do extraordinary things to their bottom line to make their annual reports look good, often to the detriment of the future viability of the company. A good example is a situation which occurred to me recently. I had been invited to give a training presentation at a conference. They said that they would pay my expenses, so I booked a non-refundable return travel ticket. Just a week and a half before the conference was due to commence, they contacted me to say it was cancelled. Their finance director had put a moratorium on travel to improve the bottom line of the company. However, since my ticket was booked, they still had to pay for the ticket, without getting any of the value of the training.

This kind of short-termism can also be seen when we look at capital budgets. In this arena, short-termism tends to be celebrated – in the building services sector, it is generally called 'value engineering'. The building services engineer might have proposed an energy-saving enhancement that might have a payback time of four or five years. This will be cut to bring the cost down, but the cheaper alternative will end up adding to the financial burden in the future. This

Delivering Sustainable Buildings: an industry insider's view, First Edition. Mike Malina.
© 2013 Mike Malina. Published 2013 by Blackwell Publishing Ltd.

happens all the time on projects, and at the time of writing it's becoming worse because of the down-turn in the economy which started in 2009.

There needs to be a total culture change, as sustainability is for the long term. But headlines are dominated by the ups and downs of the stock market and macro-economics of various countries' sovereign debt issues. The whole economic system, from stocks and shares to commodities, is driven by speculation and sentiment. Markets react to news, good and bad, and this has a massive effect on the whole economy. For sustainability to work, with regards to wider society and buildings, we must take a longer-term view. My observation is that this is not happening on a large scale. The finance system is still driven by speculators and large investment companies that want the quickest and largest return possible. I would have hoped that the lessons learned from the recession and market volatility would start to make people question and do things differently, but we still haven't changed our behaviours. What will hopefully focus minds is the significant pressure on energy and commodities prices, as the incentives will be greater than ever to make sure that savings are made. But to achieve this, investment will have to be made: as the old saying goes, you have to speculate to accumulate.

Therefore, market incentives such as the Green Deal and energy reduction schemes will become more commonplace and become part of the business culture. How we break the spiral of boom-and-bust is the big question. It's a case of developing culture and education strategies to make the finance professionals and business leaders realise that sustainability pays much bigger dividends in the long term than the short-term grab for profit at any cost. Unless we do this, society as we know it is doomed. Impacts on society will be felt in the next ten years, and in about thirty years we will be in deep trouble. Population growth coupled with the pressures on the availability of diminishing natural resources will escalate prices. We cannot go on with conventional economics. We need to change the system. We need to look at the whole way we value resources and we have to have that longer-term vision. How this can happen is a challenge, because conventions, psychology and government systems all uphold the status quo. To save ourselves, we would need a lot more intervention in the markets. We need a sustainable financial system to underpin everything we do. We need to break the mould. These are considerations which are beyond the control of ordinary people.

But for our own buildings, we can try to take advantage of mechanisms and technologies that will help us achieve the best results in terms of resources and long-term efficiency. From an organisational point of view, I would like to see integration of any sustainable agenda with the company's financial systems – I'tm thinking about issues such as energy-saving schemes and recycling. After all, finances and energy are measured in similar ways – both have a cyclic business model. Both have targets, which we measure and evaluate. Measuring sustainability, energy use and budgeting really do go hand in hand. If we assess everything we do with regards to energy and resource saving, and prioritise those areas, we will get a better return than we would if we were not doing anything at all. Reducing resource use is good business, very often offering greater financial returns than mainstream business activities.

In terms of financial mechanisms, I'm a strong believer in ring-fencing financial budgets for sustainable projects. In other words, all the savings, or at least a significant percentage of them, made from energy saving should be ploughed back to pump-prime more sustainability projects and save more energy. In this way, energy saving can becomes a virtuous circle. From a conventional point of view, given our current practices, this would really break the mould. We need to get away from the convention of sustainability savings being put in a different budget to fund an overspend or unrelated business activity. And here, there is no difference between a household budget and a business. Both have got bills to pay, and need to spend within their means. And even on this smaller scale, there are opportunities for everyone to make savings from sustainability. For instance, if everyone had their home's loft insulated, the effect on energy saving would be phenomenal. Many would find that they were eligible for a grant to do the work, and in any case they could expect a financial payback in two years, and then they would have cheaper bills thereafter. Having the heating on less would also create less wear and tear on the heating system, representing another saving. These opportunities can be extrapolated on a much larger scale for a business, and on an even bigger scale nationally and internationally.

Developing a way to integrate wider sustainable engineering and finance will be the big challenge. How can we define sustainable building services and buildings in a way that enables investors, developers, finance professionals, including valuers, and occupiers to measure and attribute a financial value to sustainability?

The answer must be to break the cycle of the way we measure finance conventionally. How we construct longer-term measures will be the big question. A good example of how measurement is being changed would be the introduction of carbon accounting and the CRC energy efficiency scheme (formally known as the Carbon Reduction Commitment). This involves measuring the price of carbon, which is obviously adding a completely different dimension to the conventional way of measuring finance in relation to energy consumed. So as well as paying for the energy, business is also paying a carbon tax on the use of energy. This may be extended in some form to cover the domestic energy market, but has been held off for political reasons, as it would be very unpopular at a time of significant energy price rises.

When the CRC was first introduced, the idea was geared to a performance table, where the top performers would get a rebate and gain financially in order to incentivise the process. However, many companies were shocked when the scheme changed in 2010, when the coalition government decided to keep all the payments, so that the CRC in effect became a carbon tax. Industry and building owners were up in arms. It seems that they didn't realise that the situation created an even bigger incentive to save, offset and reduce their energy consumption. The effect of the CRC seems to have been very much about concentrating people's minds on the price of carbon per ton. In this process, they seem to have forgotten about the initial cost of energy. By reducing their payment of the carbon tax, they would at the same time be reducing their energy bills by saving their energy resources, so creating a double benefit on energy use. People become obsessed with the politics and the tax, but they need to see the CRC as an example of integrating a financial mechanism to incentivise sustainability.

Conventional accounting is all about tangible assets and all the other things that go on a conventional balance sheet. This means that there are many examples of natural processes and sustainable issues that will not feature in conventional finance and economics. An example would be building a new road. You can quantify how much it will cost to buy the land, and similarly the cost of construction, but how do you put a value on the original green space which will be destroyed, or on the wider ecological issues. A green space has a natural value in sustainability terms, but how do we convert that into finance?

We need alternative economic measures. We need to be able to put a value on that green space. In the same way, from a sustainable building services point of view, we need other mechanisms to value wider natural resources and the impact on the existing environment and how they are integrated into building services. An example of this would be to look at the lifecycle of a building. Conventionally, we focus on the cost of the project, on how much it will cost to build with regards to labour and materials. To some degree, we are also hopefully starting to look at longer-term maintenance and energy costs, and ultimately at decommissioning and the end of the lifecycle. Still, there are so many other elements that could be brought into how we measure the value of the building, and how we run it. To cite some of the examples given previously, we could look at the productivity of a workforce within a building, and explore the value that a well-commissioned and maintained building has for the business process of the occupants. (Commissioning is discussed in more detail in Chapter 11.) These are quantifiable, measurable benefits, and we need to explore them.

Funding for sustainable building projects

Finding the finance in difficult economic times can be a challenge as well as attitudes to payback. Many companies seem not to want to borrow money to fund what others would consider a good return on investment (ROI). Especially if the debt charges would cover the ongoing cost of the loan repayments, gained from the energy savings. This is very much a convention in finance where the funds may be available but the company doesn't want to accept any risks.

There are a number of established ways that companies appraise investment potential, but many are locked into conventional attitudes and fixed ways of working as regards payback, in terms of purely how long it will take to break even. With current investment rates and financial uncertainties it surprises me that so many good energy-efficiency projects are held back because even with ROI within five years (which can cover a whole range of energy-saving technologies), these projects do not get approval for funding. All sorts

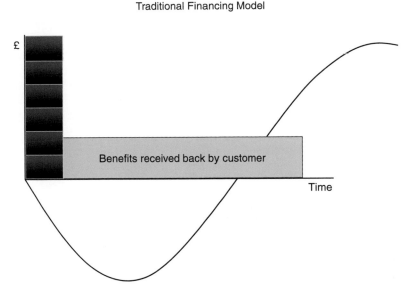

Figure 4.1 Traditional Financing Model (credit: Richard Brown, Siemens Financial Services)

of accounting techniques are used to measure traditional financial models. All rely on a capital sum to finance the project, which means that many projects will need the total cost as up-front capital. This can be illustrated as a graph (Figure 4.1) showing traditional ROI. An alternative model (Figure 4.2) develops a system of financing a project that delivers a much shorter payback – ROI funded in stages by a finance provider. This commercial model is potentially similar to how the Green Deal may be financed over the term of many projects, small and larger scale.

We also need to consider and build on the development of corporate social responsibility (CSR). There is a trend with business development to bring the wider issues of CSR into a financial accounting mechanism, often referred to as the triple bottom line: environmental sustainability, social sustainability and financial sustainability. This is an example of how an additional measurement can be brought into conventional accounting. CSR caught on because businesses became aware of the need not to incur unexpected costs in the future through their unintended negative impacts on the wider society. Modern business is very aware of the importance of public perception and reputation, and the potential impact on share price (and of course the cost of potential legal challenges) which can come from negative incidences such as pollution. This would have an impact on all three elements of CSR, thus damaging the company as a consequence.

To achieve an integration of finance and sustainability, and break down the barriers that currently exist, I believe that there has to be a significant and serious discussion between financial professional bodies and their equivalents in the engineering and sustainability field to create a mutual understanding of the issues. Experience shows that, currently, finance people often don't really

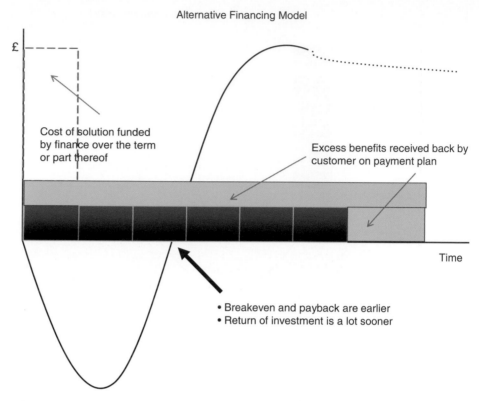

Alternative Financing Model

£

Cost of solution funded by finance over the term or part thereof

Excess benefits received back by customer on payment plan

Time

• Breakeven and payback are earlier
• Return of investment is a lot sooner

Figure 4.2 Alternative Financing Model (credit: Richard Brown, Siemens Financial Services)

understand sustainability, while engineering/sustainability practitioners don't have a grasp of finance. Once we achieve the equilibrium of finance and sustainability then this integration can easily become the norm. There are examples being developed, of how sustainability can have a wider impact on financial issues when it comes to buildings. There is a direct correlation between energy performance and the building estate from a management perspective. This is demonstrated by the desire for property portfolio managers to improve the energy rating and certification of their buildings. If the building has a good energy rating, then it will add value to the building and increase its marketability, as people want to buy or lease an efficient building.

The Royal Institution of Chartered Surveyors (RICS) refers to the impacts of sustainability on the property lifecycle. In *Surveying sustainability: a short guide for the property professional* (Fisher et al. 2007), the important point is made about linking the promotion of sustainable design, development and construction practices, including whole-life costing. Surveyors have an important role on valuations and they should also take account of the five capital assets of sustainability (natural, social, manufactured, financial and human). I believe that this type of approach will become a much more important part of the economics of sustainable construction and refurbishment in the future. Basically, sustainable building construction makes good economic and business sense.

The cheapest kilowatt-hour is the one you never use!

Reduction in energy use should be the priority issue, although many people and companies are looking at the installation of low carbon and renewable technologies as well.

The big dilemma that people are facing is when exactly to make the decision to install renewables, as there is so much conflicting information on their performance, and regarding the finance and returns on different technologies. What is clear is that energy prices will inevitably continue to rise over the longer term.

On a smaller scale, a similar question might be when to buy a computer or upgrade. A computer costs less, but it's fundamentally the same question. It's only a relatively short time since computers cost thousands of pounds. Their price has since fallen drastically and the technology has improved considerably. First-generation personal computers were large desktops and very large laptops, and they were extremely expensive. Ten years later they were smaller, and boasted far more storage space and processing power, yet they'd fallen in price to a few hundred pounds. Mobile phones have also followed a similar pattern. Therefore, we have all had to balance our immediate needs against the potential advantages of waiting for better technologies and prices. From a work point of view, we might well need a computer to function, so at some point we will have to make the choice to purchase. Still, this will be done with some regret, as we are aware that the technology will move on. Of course, the best thing to do is to buy a machine that is upgradable. I've done this on many occasions, changing the memory and hard disk to keep pace with technology. Ideally, this would be the way to go for a whole range of electronic items, renewable and low carbon technologies, to make them upgradable.

However, this could be very difficult with photovoltaics, for instance. They are composed of just one unit on the roof, a panel, and they are manufactured and sealed. These then have to be in use for 25 years, so it would be very hard to make them upgradable. During those 25 years, technology will undoubtedly develop, and future panels will become more efficient. At present, the conversion rate from sunlight to electricity is just 12–18%, but there is the possibility that this will have increased to about 25% in a few years. This will improve still further as time goes on.

Faced with this fact, the incentive for early adopters is the feed-in tariff (FIT). That is really the only incentive to invest now. Without this, it would be totally uneconomic to buy now. Those that took up the FIT in 2010–12 were guaranteed 43 pence per unit generated. The early adopters had to use the level of technology that was available at the time. If they had waited, however, in the same way that one might wait to upgrade a computer, then they would have received a diminished FIT, so it was clearly designed as a reward for early adopters, which of course would stimulate and create a market. On the other hand, as the technology develops and becomes more efficient, the photovoltaic panels will generate more electricity as the conversion ratio from sunlight to electricity will have improved.

Currently, much of the attention of the media and fashion is on photovoltaics, because they are in vogue and attract the FIT as a significant financial subsidy. We tend to like new technologies and want to be part of their development in society. Nevertheless, this can distract our attention from the fact that the cheapest kilowatt-hour of power is not solar: it's the one we never use in the first place. I would maintain that the key is still to focus on the energy hierarchy, since the financial and environmental impact is far better served by reducing the energy in the first place. See the Introduction and several references throughout this book for more on the energy hierarchy. Also, keep control of the energy that you are using. Perhaps another way of looking at this is to combine hierarchy with finance.

Whilst reducing usage of energy is key, that is not to suggest that I'tm not keen to see further investment in renewable energy. Once we'tve achieved the first steps on the energy hierarchy we must ultimately invest in renewables, perhaps by ring-fencing the savings from energy use reduction. That way, we could fund our photovoltaic panels from the energy hierarchy savings made earlier by reducing the energy demand.

Finance needs to be coupled with a measurement of the energy performance and lifecycle of the entire project, as mentioned before. One idea put forward is the concept of the primary energy ratio. This is the relationship between the amount of primary energy (for example, fuel) used and the amount of energy delivered to the end user. For example, how much energy does a boiler or a heat pump need in order to create a certain amount of hot water for delivery as useful heat? In terms of a heat pump, the coefficient of performance (COP) is defined as the ratio of heat delivered by the heat pump and the electricity supplied to the compressor. So, 1 kW of electricity can give 3 kW of heat, representing a COP of three, which can be viewed as 300% efficient. If this is related to finance, then that looks like a favourable financial ROI.

This needs to be developed further not just by looking at the energy efficiency of the energy delivered, but by ensuring that it is evaluated for its entire lifecycle, including the embodied energy from manufacturing. In other words, we have to look at energy and all the other resource issues and impacts from cradle to grave for all activities we employ inside and outside our buildings and all the associated activities. Most importantly, there needs to be a simple process for the end user and/or consumer to evaluate and understand the true impact both financially and from a sustainability impact. I propose that we adopt the tried and tested A to G rating method, which gives everyone a good understanding of what is efficient and what isn't. I've always maintained that the complex arrangements for making sure that a universal and robust system is put in place can take place behind the scenes. For the public and end user or decision maker, the simpler the system the better.

A–G can be used universally to label all products on a common measurable scale of energy with the cost and lifecycle data. This is developed in further detail, with regard to energy performance, in Chapter 6, and was also referred to in Chapter 1 as I believe the mechanism is in place to develop a universal application of this rating.

What is the Green Deal?

The Green Deal was introduced by the coalition government in 2010 and was billed as the government's flagship environmental policy. The first stage of it is aimed at domestic homeowners, to give them an incentive for major energy improvements to houses. The ultimate aim is to extend the scheme into business and the public sector to improve their building stock, and thus 'green' those sectors as well. The scheme has been created as part of the Energy Act 2011 and lays down a new financial framework to enable the delivery of fixed improvements to the energy efficiency of domestic and non-domestic properties, funded by an additional charge on energy bills that avoids the need for consumers to pay upfront costs (Figure 4.3). The Green Deal finance scheme has a 'Golden Rule' to be satisfied, with the assessment providing the basis for whether the predicted savings made by energy efficiency improvements to a property will be equal or greater than the installation cost of these improvements (DECC 2011).

The Green Deal works through a package of energy-efficiency measures, which is funded with no upfront costs to the homeowner by a Green Deal provider. The cost of the measures is paid back over a term of 25 years, through repayments on the consumer's energy bills, whereby the savings go back to the scheme. The way that this is measured is by the energy supplier providing the information so that the Green Deal Provider can calculate the repayments. The cost of the repayments is less than or equal to the likely energy bill savings. The difference with this financial package from the structure of a conventional loan is that the customer will not be liable for the capital sum.

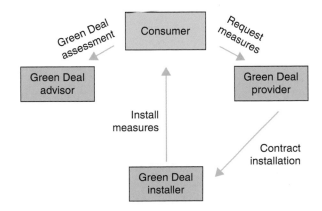

Figure 4.3 The Green Deal finance process

The repayments are only made while the homeowners remain in the property. When they move or when the property changes hands, the obligation to pay the Green Deal provider will pass on to the new occupier. This will include owner occupiers, private social rented sector and the industrial and commercial sector.

This whole process is initiated and assessed by Green Deal assessors, so a whole new section of the industry is being created. This provides opportunities for contractors and energy management specialists to add to their portfolio of business opportunities. The assessment and the advice is based on an extension of the energy performance certificates, which are discussed in Chapter 6 of this book, but encompasses a lot more new data and parameters which will have to be factored in, such as the way the building is being used by the current occupants. This won't work without energy use and behaviour being studied – another new dimension of this scheme. I view this as positive, and support the move towards measuring usage and behaviour as well as buildings. It's time we were positively addressing the human factor and looking seriously at how to integrate both the technical and the people sides of the equations. As an industry, construction tends to be concerned with technology and number crunching. It doesn't routinely address the people issues. We urgently need to establish processes to change this.

How does the Green Deal work? To obtain the finance, projects have to be assessed for eligibility (Figure 4.4). Therefore, there need to be issues in the building which need to be addressed, in order for everything done to be eligible. At the time of writing, however, it seems likely that the bulk of buildings will be eligible. The next part of the process is the physical assessment measurement, which is carried out to establish which measures are the best options to be installed, and to assess what has already happened. Is there already loft installation, for instance? If there's room for more, it would be good to install it, but obviously this would not be proposed for a house which already had the recommended maximum amount. From a survey of each property, the Green Deal assessor will make recommendations on which measures should be retrofitted or installed. These measures will then be assessed to establish which are suitable to be funded.

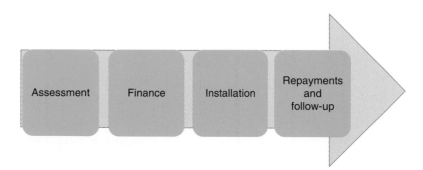

Figure 4.4 Green Deal process cycle for Installation

The best possible way of doing this will be by using the energy hierarchy and calculating the likelihood of payback from each proposed measure. This will be a combination of the cost of the efficiency measures themselves combined with the cost of installation. The final cost will have to be assessed against the likely ROI. For the scheme to work, an intervention has to pay back in 25 years. The type of measures that will be eligible will be a range of heating ventilation and air conditioning, lighting, water heating and building fabric.

Ultimately, any energy efficiency measure to improve building performance which can be financed from the saving on the energy bills should be eligible. The key will be the financial package and how it's employed, because

The following measures are likely to be eligible for the Green Deal (DECC 2011).

Heating, ventilation and air conditioning
Heating controls
Air source heat pumps
Ground source heat pumps
Condensing boilers
Heat exchanging/recovery systems
Mechanical ventilation (non-domestic)
Flue gas heat recovery devices

Building fabric
Cavity wall insulation
Loft insulation
Flat roof insulation
Internal wall insulation
External wall insulation
Draught proofing
Floor insulation
Heating system insulation (cylinder, pipes)
Energy efficient glazing and doors

Lighting
Lighting fittings (non-domestic)
Lighting controls (non-domestic)

Water heating and efficiency
Water efficient taps and showers
Innovative hot water systems

Micro-generation
Solar PV
Solar thermal
Biomass boilers
Micro-CHP

the finance will be tied in with how the energy is measured via the fuel gas and electricity meter. The consumer will be responsible for repayments as part of their lives as they consume energy. Nearly all of this Green Deal will be financed via the private sector, so it's likely that large consumer outlets such as the big DIY stores and utility companies and perhaps some banks will be the private sector finance providers.

A whole raft of legislation is being amended to cover this change. There will need to be changes to the Consumer Credit Act, to debit collection, to name but two, and issues around vulnerable consumers are also having to be addressed.

The guidelines make it clear that the Green Deal provider must give appropriate advice within the terms of the Consumer Credit Act. This is vital and must be enforced as the prospect of another miss-selling scandal would be calamitous to making this whole scheme work. It would also damage the vital effort of making energy efficiency popular with the need to make people much more energy aware.

A whole series of accreditations are being established so that the Green Deal assessors and installers can be trained and approved to access this scheme. The key for the contractor will be the installation stage. The technology will have to be approved and assessed, and there will have to be an approved installer's regime which will hopefully kill the cowboys, as I state in Chapter 13.

Ultimately, there will be a whole new regulatory regime and Green Deal quality mark. New technological standards will also be implemented; for example the British Standards Institute is developing a publicly available specification covering the certification elements of the products and materials and the technical criteria for the physical installation of the products and materials. There will be an assessment of the installer's technical qualifications, together with a whole series of measures that cover the consumer process, complaints and warranties. Ultimately, the whole of this accreditation process will have to be independently accredited by the UK Accreditation Service (UKAS). This is going to be stringent, so it really will be a significant blow to the cowboys that have dogged the construction industry to date.

The importance of accredited installers and the necessary regime for quality control and enforcement of standards cannot be over emphasised. The Department of Energy and Climate Change in a departmental press release, (DECC 2010) stated 'the Green Deal is a massive business opportunity which has the potential to create up to a quarter of a million jobs.' The then Secretary of State called this part of the third industrial revolution. He was thinking about insulation installers and the supply chain around that as well as Green Deal assessors, so his prediction may be accurate. The big question for the whole Green Deal is just how much consumer confidence can be established in the practice. There will be a number of operational questions – will people be happy to buy property with another loan against it? But as the householders are not liable for the capital payments, this arrangement may just become the norm. This scheme is due for launch in October 2012 for the domestic sector and will be followed by the commercial sector in 2013. But, at the time of writing this timetable looks likely to slip. The scheme does have great potential if the government can get the mechanism right, with the scheme and access to funding not becoming too bureaucratic. There is no doubt that there is a vital need for a catalyst to stimulate a significant uptake in refurbishing the very large numbers of buildings across the UK to make them a lot more energy efficient.

References

DECC (2010) Department press release 10/104, 21 September 2010 http://www.decc. gov.uk/en/content/cms/news/pn10_104/pn10_104.aspx (accessed 15.8.2012)

DECC (2011) What measures does the Green Deal cover?Available at: http://www. decc.gov.uk/assets/decc/what%20we%20do/supporting%20consumers/green_ deal/1734-what-measures-does-the-green-deal-cover.pdf (accessed 15.8.2012)

Fisher et al. (2007) *Surveying sustainability: a short guide for the property professional*, June 2007, RICS Royal Institution of Chartered Surveyors

2 Delivering sustainable buildings

5 Delivering an energy-efficient and sustainable building

Designing and constructing a sustainable building is not a simple task, but it should not be seen as any more complicated than a 'conventional' building that has characterised the recent past and present. Sustainable buildings are the future and will, over the next few decades, be more commonplace. The construction industry is going through a revolution through the integration of working methods and evolving technologies, as the construction and refurbishment of low carbon buildings gathers pace.

Doing the job properly involves a wide range of professionals. From my own experience, having worked on a number of projects, the level of success of the project largely depends on the standard of planning and the ability of the team to communicate. If everything's mapped out, roles are clear and things are explained thoroughly, then the project will go well, but there needs to be a framework for everyone to refer to.

The design and construction phase has traditionally been broken down into a project plan, which is crucial, because the design of the plan itself is what sets up the terms of reference for the whole project. The common denominators of all projects are time management and a physical sequence of events which has to be grounded in the achievable. It has to be a defined process. Something that I think has been beneficial is where project teams have an opportunity to spend some time gelling as a group. It's particularly useful if participants get a chance to talk through their role on the project, so that everyone understands everyone else's task. The commissioning manager can sometimes facilitate this – in my experience it is not uncommon for them to take on some of the project management tasks.

I have recently had the opportunity to put this belief into practice in a wider forum, as it's been my privilege to chair the Sustainable Built Environment East group. This is a unique group in the East of England that has brought together architects, surveyors, town and transport planners, civil engineers, building services engineers, main contractors, mechanical and electrical subcontractors, NGOs, local government and, until abolished, regional government. Everyone involved has a shared interest in agreeing the way forward to achieve a sustainable built environment. This has not been project-specific but has begun an important process in creating a much better understanding of the different disciplines and specific roles in the sustainable built environment. By using my philosophy of the importance of creating a shared understanding of professional roles, this has done is what I predicted – it's reinforced the collective idea of what we're trying to achieve and has been useful in reminding people how many disciplines and expert roles there are in this field. It's created a massive network within the eastern region, including crossover with the local Construction Industry Council, employers, colleges and further education colleges, as well as developing links with SummitSkills and ConstructionSkills, the construction and building services sector's skills councils. It's always been a solid networking group, but I believe I've strengthened this by placing special emphasis on the shared understanding of roles and tasks.

The process of sharing and mapping should be integrated with the process of building information modelling (see the Introduction for more on this). A combination of modelling, documented systems and positive team communication should be the way a successful sustainable building is constructed and refurbished on a larger scale in the future.

If this shared understanding is not fostered, then we, as an industry, have a tendency not to fully appreciate the contribution of others, creating unnecessary industry divisions. Construction can be tribal, and this can produce a tendency to blame other people when things go wrong.

The positive counter to that is that if it's clearly laid out in the beginning and everyone feels part of the process, it makes life so much easier and the project team far more effective, if everyone feels listened to and valued. You get more out of people if they feel appreciated. Communication is key. It's also important to remember, as recent commentators have noted, that '…effective communication has two components. We don't just need to be able to present information effectively, we also need to listen, and question where necessary to establish that we understand what we've heard' (Sullivan et al. 2010).

The wider design process – BREEAM

If you ask various architects about design, you may get very different answers from different architects, but you will also get a very different idea about what a given design is trying to achieve than if you spoke to, say, a surveyor. Ideally, you want the architect to be a master of all trades (although these gifted individuals are, sadly, very few and far between), at least in the sense that they can appreciate the role of all disciplines throughout the building process. An architect, by profession, is trained to look at functionality, design and aesthetics, but may not see the infrastructure and how the building will physically function from a building services perspective. But if they don't appreciate building services, and its challenges and constraints, it will impact on the sustainability and functionality of the building.

There are many questions to be considered in the wider context of delivering a low carbon sustainable building – unfortunately, in the past, many architects have not asked them. Where is the electricity supply coming from, for example? Is it from a sustainable source or the mix of supply? What are the implications on water resources? What is feasible for goods movement and public transportation? Does the general infrastructure fit in with wider town planning? These issues have been discussed to an extent in the Chapter 2, which considers planning in more detail, but everything in construction is related, and they are of course also integral to design.

One of the ways of answering these questions is through the influence and input of the Building Research Establishment Environmental Assessment Method (BREEAM). There is also a US version called Leadership in Energy and Environmental Design (LEED) which can also be used. What BREEAM does is to provide a standard, and one day it, or something like it, may be incorporated into legislation, which I believe would be a positive move towards fostering truly sustainable integration in the built environment.

So far, we've talked about all the different professions and some considerations that wouldn't have normally gone into a building process, because each discipline is mainly interested in their own part of the structure. The BREEAM method, or LEED standard, brings in those wider built environment and sustainability issues.

BREEAM ratings can be given as a pass at the lowest level to outstanding at the highest level. To achieve the ratings, a series of compulsory standards must be met as a minimum. Once these standards have been reached the project will be able to gain additional credit points to obtain a higher rating. Table 5.1 shows the BREEAM ratings (BREEAM 2011).

To get a high BREEAM rating, you need extra points for accessibility and waste management, as well as for the construction process itself. Each specialist can contribute expertise towards getting the maximum number of points to achieving an excellent or outstanding building, so collaboration becomes part

What is BREEAM?

BREEAM was launched in 1990 and developed by the Building Research Establishment (BRE). BREEAM is an environmental assessment method for new and existing buildings which is based on a sustainability rating determined by a series of credit points. The standard gives best practice in sustainable building design, construction and operation of a building and has become one of the most widely used and recognised measures of a building's environmental performance.

A BREEAM assessment uses recognised measures of performance, which are set against established benchmarks, to evaluate a building's specification, design, construction and use. The measures employed represent a broad range of classifications and benchmarks ranging from energy to ecosystems and natural environment. This includes issues related to energy and water use, the building internal environment and indoor climate (health and well-being), materials, waste, transport, pollution, ecology and wider management processes.

Further information from: www.breeam.org

Table 5.1 BREEAM credit points score to achieve building rating

BREEAM rating	% score
Outstanding	≥85
Excellent	≥70
Very good	≥55
Good	≥45
Pass	≥30

of the early stage process. The BREEAM process is even more important because you get more credits for BREEAM if you do things earlier in the Royal Institute of British Architects (RIBA) work stage process, so the incentive shows itself even more. If, for example, you wanted more points for automation and building controls, the earlier you adopted these measures in the design stage the more points you would get. It won't work if building controls are left as an afterthought. Therefore, a vital component of a building such as controls should be thought through as early as possible.

Learning from the manufacturing and car industry

I've often thought that just as in industrial and manufacturing processes, where you have a complete parts list, the same should be done for buildings. This is actually being used in offsite construction projects now, but needs to become more widely adopted. In the construction industry we have quantity surveyors, whose job is to be 'the construction industry's economist and (who) manages and controls costs within projects, involving the use of a variety of management procedures and technical measurement tools' (RICS 2012). But they often work in isolation or specifically to the subcontracted roles within the construction process.

If this was coordinated and planned, it would provide all the data and records for an enhanced building log book (Chapter 11 has a fuller discussion of log books). This could also be an extension supplied by a BIM system (see Introduction).

If a parts list was part of designing the building and actually constructing it, I believe that the industry and legislation could move one step further into this fairly complex area and start to plan out the lifecycle and sequence of embodied energy before construction begins. In effect, what I think will happen at some stage is that if buildings have complete parts list then each component could have its embodied energy content listed. This sounds unwieldy, but with currently available IT, it's very possible, and thus we could end up with a true carbon footprint based on detailed, accurate data. This would give a true and entire lifecycle prediction of the carbon footprint of a building.

This would be more meaningful, in my view, than any existing technique as a measure of how sustainable a building is.

Designing for sustainable communities

To the traditional building services engineer or project manager, all of these wider considerations would not often seem a priority in their day-to-day working life due to the nature of their specific jobs roles. However, we will all need to change to embrace techniques and practices that work towards achieving sustainable buildings. We will have to learn and adapt and get used to talking about linking all the issues together. This will also be expected as part of the planning process, and of all the building control processes and statutory requirements that need to be fulfilled to proceed with the building itself. These requirements would include issues of biodiversity, involving professionals such as landscape architects and also consultations with local wildlife trusts and groups, even things like habitat management. A lot of these issues will have an impact on the actual building. A greener environment has been shown to create a more pleasant environment for all. Think about what type of planting the development should include. The plan should encompass indigenous plants and should encourage an environment where the biodiversity of wildlife can be maximised. Additional issues such as economic integration and planning sustainable economic developments, which includes local employment, will have to be considered. This has links, of course, with transportation, because obviously if locally employed people don't have to travel so far, there is less environmental impact. All of these issues would be considered as part of the planning process (see Chapter 2) and are therefore integral to the design process.

In the design process, you almost have to be able to look into a crystal ball – how is technology going to change? How is its use going to change? There are issues of space planning and of looking at a building in terms of its subdivisions and modularity. The best buildings have the best adaptability. An example of how this can be achieved is the concept of demountable partitions. Some organisations have a high churn rate, which impacts on space use. There's been a trend in the past to move from open plan offices to modular offices and back again. New partitions used to be put up without services being thought about. Therefore, some offices ended up with an extract of air but no input of air! Other examples of such bad planning include thermostats and light switches being outside of the area they control. All of these things need to be thought out, and the advent of technological change (plug-and-connect systems) means that you don't have to permanently hardwire many of the buildings services; you can just plug them in. The same applies to building controls. With the advent of wireless controls, it makes it easier to move the sensors to adapt to the way the building is being used over time.

During the evolution of planning and building regulations, accessibility has become a major issue. Accessibility will need to be created and maintained. All of this is covered in Part M of the Building Regulations (see Chapter 3).

To achieve the best possible sustainable design and development, all resources and impacts need to be considered in a wide ranging and holistic way to achieve the lowest possible use of resources and energy during the construction and lifecycle of the building and the surrounding environment.

All resources need to be thought about and, as mentioned earlier, the latest BREEAM standard takes account of waste, through the application of benchmarks for predicting and forecasting construction waste. Water and transport are also taken into account. Site waste management plans are also a legal requirement for large construction projects under the Site Waste Management Plans Regulations 2008.

There are two forms of waste in construction: waste from the construction process itself and waste within the lifecycle of the building. The component parts contain embodied energy, and the embodied energy of the waste materials needs to form part of that calculation. There will always be some waste in construction, but obviously every effort has to be made to minimise this. We always need to bear in mind the waste resources hierarchy (see Chapters 8 and 12, especially Figure 8.3). Best of all is the reduction of waste. The next best solution is reuse of the materials, then recycling. The worst solution is disposal, so we should be working towards eliminating this element as much as possible. A useful free tool is the SMART Waste Plan, which is produced by the BRE. This has been designed to assist in mapping out a site waste management plan and as a waste measurement tool. It is also useful as a tool to help meet the requirements for BREEAM credits for commercial projects and the requirements of the Code for Sustainable Homes standards for houses.

The second kind of waste, that occurs during the operation of the building, again needs to be addressed bearing the same hierarchy in mind. Design of the building itself is an important element here. It's desirable to include storage and sorting facilities for the waste. Recycling bins are often an afterthought, but proper design would include access for easy movement, collecting and transporting of these facilities. Ultimately, in a larger development, there should be shared resources. Places such as industrial estates are a good opportunity for this. It would be good at the design stage to consider whether such facilities could also be for community use. After all, sustainability is not just about individual buildings, but about shared resources for businesses and local people. These kinds of solutions would likely be very favourably received by the local authority planners. It's not just about ecological altruism – this kind of initiative will almost certainly tick a lot of boxes at the initial approval stage!

The same applies to water. Water will now become a major planning and strategic issue for local and water authorities (see Chapter 7). In terms of design, there will be a lot more emphasis placed on water use in construction. As there is a new emphasis with regards to the calculation of embodied energy, part of this equation will be measuring how much water goes into manufacturing of materials and also how much is used during the construction process. During the operation of the building, we will need to consider issues around minimising water use, with strategies such as reusing water through rainwater harvesting and collection, using this water to flush toilets, for example, and reducing the usage of treated water supplies. There will also be the possibility of using grey water recycling, through the recycling of wastewater from sinks etc. There are some buildings which will even incorporate what is known as black water, using natural filtering from the ecology of reed beds. This will be less common in urban areas, but may be feasible in highly rural, well-managed sites.

Transport is another issue. It should be remembered that there are points in BREEAM for transport. You will also get extra points for local planning if you design a building with facilities for cyclists, who will need both shower and storage facilities. We need to think about maximising the availability of local public transport, through pedestrian access and bus stops. With regard to car parking, there has been talk of taxing people's car parking spaces. There will be more financial incentives in future – Nottingham City Council introduced a 'workforce parking levy' on firms with more than ten staff car parking spaces in 2012, and other cities are said to be interested in joining in. In a climate such as this, shared car pooling schemes need to be considered by all organisations. These schemes need to be planned for in advance in the design of the building. The other issue with extensive car parking spaces is that land is also expensive.

Where is the electricity coming from?

One of the dilemmas in sustainable building design is deciding where your electricity is going to be sourced from. You could have the greenest building on earth in terms of how you manage and control building and energy use, but where is that energy actually coming from? There is a wide debate about the future of electricity supply. Most of the electricity in the UK currently comes from fossil fuels, mainly coal and gas. A breakdown is given in Figure 5.1.

There is an opportunity for developers and building operators to purchase a greener mix of electrical supply. Of course, this doesn't mean it's actually a green source, as whatever power you buy comes from the grid, but in terms of the energy mix, you could specify that the source of power you are purchasing is derived from a low carbon source such as hydroelectric, wind or, perhaps in the near future, tidal power. There is a debate about nuclear power. Some say it's low carbon and green, others would say that this is wrong due to its embodied energy, and that it's not sustainable, since uranium is a finite resource and has severe implications for the future with regards to how to deal with nuclear waste (this is mentioned in Chapter 1 as part of the sustainable dilemma).

Because it's difficult to truly obtain a green source of energy without a locally based wind farm or hydroelectric station that you can draw from, you

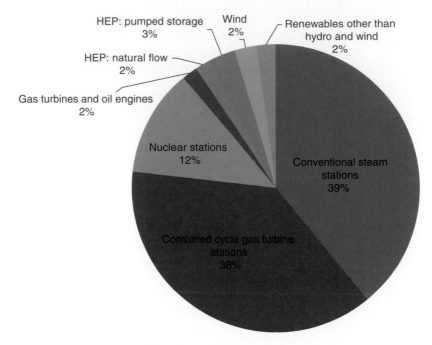

Figure 5.1 UK electricity supply breakdown 2010 (Dukes 2011)

don't have a lot of choice and control over what you receive. The other possibility is to influence the efficiency of the grid to reduce demand by everyone adopting dynamic demand control, which in my view should be mandatory.

Dynamic demand control

One positive development towards mass scale influence on of the delivered electrical supply to the building that should be adopted is the concept of dynamic demand control. I believe that eventually this concept will be integral to the way buildings behave in relation to managing supply and demand of their electrical energy. How it works is as follows: The UK's power providers and most other electricity networks have to supply a base-load of power. A base-load is a supply that is always there, and always running, to adjust for the eventualities of supply and demand. We are aware that there are peaks in demand – for example, when there is a very popular TV programme on, such as a crucial world cup match, there will be a significant peak in demand at half time, when millions of people get up at the same time to put the kettle on. A million three-kilowatt kettles (3 gigawatts) represents most of the generating capacity of western Europe's largest power station, at Drax in Yorkshire.

What is a watt

Electrical power is measured in watts (W). Named after Scottish inventor and engineer James Watt. A single watt is a small amount of power, but soon builds up to a much larger amount used and consumed in domestic and larger commercial situations. The electricity bill you receive details energy consumption in kilowatt-hours (1 kW = 1000 watts). The electricity meter, shows how many kilowatt-hours (or 'units') have been consumed. One kilowatt hour (kWh) is the amount of energy used when 1000 watts is consumed for one hour.

An example is a conventional 100 W light. This is a tenth of a kilowatt, so it will take ten hours to use a unit or 1kWh. A kettle rated at 3 kW if left running for 20 minutes equivalent would use one unit or 1kWh.

The cost of a kilowatt hour (unit) varies by electrical supplier and by geographical area. This can also be complicated because some utility companies charge different rates for different number of units used, as well as applying other fixed supply charges.

Power and energy are frequently mixed up: power is the rate at which energy is generated or consumed. Power is measured in watts. A unit of energy is the actual energy used measured kilowatt-hours.

When it comes to larger measurements of usage, or the capacity of a power station, the units of megawatt or gigawatt is used.

kilowatt	1 kW = 1000 W
megawatt	1 MW = 1000 kW
gigawatt	1 GW = 1000 MW

So power stations are measured in MW or GW of electrical power output. for example, the Drax coal-fired power station can generate up to 3870 MW which represents 7% of the UK electricity supply.

Electricity can't be stored in significant quantity, so suppliers have to provide a constant base load, or 'spinning reserve', by keeping enough generators rotating to allow for these fluctuations in demand. These peaks have to be predicted – get it wrong, and the outcome can be power cuts. The industry tries to predict these peaks as accurately as it can, but they have to have a margin for error and that means waste.

However, there is a way that we can employ technology to help reduce the need to keep generators spinning without generating electricity thereby wasting energy. On a large scale, we don't have a means of storing the energy, but we can use dynamic demand management or control. This would use devices which could be fitted to a range of electrical appliances, which would measure the supply and demand level on the national grid. These appliances would have to be non-critical electrical devices, such as fridges, freezers and air conditioning units. They could be any device where it wouldn't matter or even be noticeable if it was powered off for, say, 10 minutes. The technology would be controlled, so that appliances won't loose power for a long time, but just 10 minutes would be enough to get control of the base load on the grid.

There's been a lot of debate about this technology, about getting this put into legislation to make it a mandatory item. It seems like common sense, so why hasn't it happened? Well, it's been debated in parliament, but in my opinion, there are a lot of vested interests, namely the power companies, who are paid significantly to maintain the spinning reserve and have those generators on standby.

In California, New York and parts of Canada, however, dynamic demand control is now becoming a reality. The catalyst for California is that it ran out of power during recent heat waves, due to massive demand for air conditioning, so there it's been a useful solution. It is one of the main things that we as an industry should be helping to fulfil, in order to reduce carbon emissions. If every building had these devices fitted, they could contribute to making this a reality. It would certainly create a more sustainable building. But while there has been a lot of lobbying for legislation, in my opinion, the UK government of all parties has no political will to create legislation and take on the electricity supply industry. I believe that the only possibility of this becoming legislation will be if there is a directive from the EU. The problem there, however, will be the different legislative control arrangements that different countries have over their power supply. Still, I think this is something that the industry should make a priority, and we have campaigned for this through CIBSE. Ultimately, what we want is a more efficient and stable electricity grid, and it would help reduce the cost of lower carbon and renewable energy.

Looking ahead

In Chapters 6–12, I will cover all the dimensions which will be needed for a successful BREEAM rating, as well as other considerations which can lead to an even more enhanced sustainability performance for a building. This together creates a holistic view of what is required for delivering sustainable buildings. In Chapter 6, I cover the details of an energy-efficient building, and how standards and energy ratings have evolved. We will also consider how they will improve the process of construction. In Chapter 7, I concentrate on water and its importance. It's now being referred to as 'the new oil', and it will gather more and more attention as time goes on. Chapter 8 focuses on the role of the contractor, both their importance in delivering the physical building and the building services engineering. I also look at the business opportunities for contractors which will arise from the evolving sustainable buildings market. Chapter 9 deals with the traditional main components of building services engineering – HVAC systems – with a consideration of the sustainability dimension. Chapter 10 considers the vital importance of the brains of the building, and shows how traditional building management systems can truly become a building energy management system, thus resulting in real systems integration. Chapter 11 looks at the importance of the continuous commissioning process, built on a sound foundation of a proper commissioning programme throughout the entire building process from cradle to grave. This leads on to Chapter 12 and the importance of a positive regime for planned preventative maintenance to enable a sustainable building to be kept functional and efficient, and keeping it all going.

References

BREEAM (2011) *Technical Manual*, BRE

Dukes (2011) Electricity Statistics Chapter 5: Tables 28th July 2011 http://www.decc.gov.uk/en/content/cms/statistics/energy_stats/source/electricity/electricity.aspx (accessed 15.8.2012)

RICS http://www.rics.org/site/scripts/documents_info.aspx?documentID=565 (accessed 15.8.2012)

Sullivan, G. et al., 2010, *Managing Construction Logistics*, Wiley-Blackwell

6 Managing energy and reducing its use

Fundamental to making sustainable buildings a reality is the process of energy management and efficiency. To date, this has been mainly done by either facilities managers or, in more enlightened organisations and many larger organisations, dedicated energy managers. Whilst energy managers are currently in the minority, the role is not a new profession; it's actually been around for a long time.

Energy is too cheap

However, if we look back to the 1970s and 1980s, the role was often seen as not about managing energy use per se, but about reducing costs and managing supply tariffs. The energy manager's role started to make its presence felt in the first oil crisis of 1973–74, when petrol looked like being rationed, and there were real concerns about the Middle East and the blockade strangling fuel supply. The government took the situation seriously and talked about energy-saving campaigns. A number of organisations then had dedicated professionals looking solely at this issue. This was a good start, but then as the oil blockade was lifted, the role fell into a malaise, because, aside from fuel supply during the blockade itself, energy was still relatively cheap.

This brief, promising move towards energy management was eroded further in the 1980s. For at least a decade, the privatisation of energy companies held prices down through the artificial creation of competition. The move broke up the power companies, and this drove down the price of supplies to business and eventually the domestic consumers. The market created in the 1980s has evolved since, with takeovers and buyouts creating bigger and more powerful utility companies, so, all in all, the map has changed completely over the past quarter century.

This whole process was a disaster for the energy manager, because the new role of the gas and electricity companies – essentially, as glorified brokers – meant that some companies could buy electricity at almost 50% less than they had been paying before. The 'bean counters' were more than satisfied with this, and accordingly the energy managers' budgets were cut. They had fewer resources as there were fewer incentives to save further. The underlying problem was the focus on price, rather than on the actual energy use. In the 1970s, there had actually been a move towards reducing consumption, due to the issues around security of supply and potential costs. In the 1980s and 1990s, on the other hand, the focus was all on the money. This set everything back, as there was far less incentive to save energy.

Essentially, it's taken all this time to get back to where we were. Decades of potential were lost. Only in the 21st century have prices started rising again to an extent that people notice and businesses take it more seriously. The true nature of supply and demand is starting to take effect, because of all the other issues around world economics. We are back to security of supply issues for energy and a range of other metals, mineral and food commodities. Also, in the background, there is always the supply and price of crude oil.

In real terms, however, energy is still too cheap. I should emphasise that this is not true for the fuel poor (those that spend more than 10% of their household income on fuel), but for companies, energy is still very cheap unless you're an intensive energy user. Relative to all the other costs of a business, its energy costs are small beer.

We can see from the story so far that the history of energy costs, and the history of the energy manager's role, has been full of backward steps from a sustainability perspective. Budget holders' overall bills had gone down significantly, so the return on investment in energy management went down as well. Competition can makes things cheaper, but that was bad news for the development of the energy management profession. Energy was so cheap that it had to rise significantly before it became a big issue for companies again. Even now, it's still so cheap in relative terms that other factors will be needed to motivate people – energy taxes for large companies, for instance and potentially domestic customers as well if a government has the courage to implement this.

Energy use and carbon taxes

Currently in the UK there are four energy taxes: the emissions trading scheme (ETS), the climate change levy (CCL), the carbon price floor (CPF) and the CRC energy efficiency scheme (CRC). Just as with legislation there is some confusion over the different schemes and a number of bodies and companies complain that they are over complicated.

The ETS and CPF will have major implications for energy generators and large industrial companies as well as having a big knock-on effect to all of us as the tax on carbon and energy use is increased significantly over the coming decades.

A UK Treasury consultation document (HMRC 2010) sets out three separate proposals for the CPF, and outlines how the price in 2020 will be delivered through a new tax combination with the ETS, which could reach £20, £30 or £40 per tonne. Each scenario set out price options with the price continuing to rise to reach £70 per tonne in 2030. The implications for energy use for the future and the incentives to reduce use and to reduce carbon are clear.

The CRC (formally known as the Carbon Reduction Commitment) also has a big part to play here. This scheme targets larger, less energy-intensive organisations with an incentive to reduce the level of carbon emissions by 1.2 million tonnes of CO_2 by 2020. This is part of the Climate Change Bill commitment (see Chapter 3 on legislation) to put mechanisms in place to achieve the overall targets of reducing CO_2 emissions by at least 80% by 2050.

Those companies that have an active energy-saving policy and programme will benefit from lower energy bills and will lessen the financial impact of the current £12 per tonne payable as part of the scheme. The scheme also publishes a league table, which offers another incentive for companies to aim for the highest place and gain the best reputation for their energy and environmental performance.

All sorts of mechanisms will have to be used to incentivise companies. Energy prices alone won't do it – people moan, but it's not having the impact it should. Therefore, we need a carrot and stick approach, to create a combination of incentivisation and taxation.

Energy management must be integrated

The other historical practice regarding energy managers is that they have tended to be compartmentalised. In the past, energy management was not integrated with other building services functions. For example, only recently has there been any crossover and integration between energy management and commissioning and maintenance. My own experience of promoting this integration began when working with Commtech, one of the then leading commissioning companies in Europe. This change has come about because of the effects of implementing good systems of planned preventative maintenance and other commissioning practices, which have been shown to be very effective in saving energy and reducing demand (This is developed further in Chapter 12). Full integration of energy management and other building services functions and activities is essential for all businesses.

Many energy managers have concentrated only on the 'process' – the number crunching that goes into saving hours of plant and equipment operation. The 'people' side of energy saving has largely been neglected. There really hasn't been an awful lot of effort towards changing attitudes and individual practice, apart from the old 'save it' posters, which were frankly rather bland in any case. Sticking a poster on the wall or round the light switch really isn't good enough. It just becomes part of the scenery. A lot of energy managers are going through the motions in this regard. Not until more recent times have there been more active campaigns geared to educating and persuading people to buy into the process of saving energy, and this is still a very new phenomenon.

However, we are beginning to see a trend towards more of an integration with behaviour management, with energy management almost crossing into human resources functions. With behaviourist techniques being used more, so energy managers have had to evolve and adopt a wider brief and learn from other professionals. To a large extent, this mirrors the process whereby building services engineers are becoming more generic. We can also see similarities in the way health and safety issues have evolved to become a mainstream part of day-to-day practices and processes.

The big difference between the roles is that, unlike building services engineers, the energy manager is often only appointed when a client or building owner is taking occupation of the building, which means that they've had no input into the design, construction or commissioning of the particular building. Therefore, they have to learn from scratch all of the processes and functions within an existing building. In an ideal world, forward-thinking companies would recruit the energy manager during the construction process where possible, and integrate that person within the construction team. They would then have a much deeper understanding of the building and its building services function. This links with the more advanced techniques which are beginning to be established to take account of how people will use the building. Building information modelling (BIM) can also be used to plan energy management in advance as part of the design and construction process.

The energy manager role sits in very different positions in the management structure or processes of different companies. It seems to be split between facilities management and the building services engineering department. It might even be within a corporate management unit looking at business processes. Sometimes the health, safety and environment advisor has had the energy management role tacked onto their job description. This is not ideal, because it's really a full-time job in itself! The UK Audit Commission recommended that for every £1 million spent on energy in an organisation, a full-time energy manager should be in post. So for a larger organisation, for example, with a spend of £5 million, five full-time energy management professionals should be in post. Unfortunately, this is rare. It does actually make business sense, because for a larger organisation the salary of any energy manager should easily be covered by the savings made. It is fair to say that the full-time energy management professional is more than likely to be completely self-financing.

The role of the energy manager can encompass:

- developing and implementing energy saving strategies
- ensuring compliance with EU directives and other legislation
- monitoring and reporting on energy management progress
- negotiating energy contracts
- maximising energy efficiency in the short and long term, including training of building engineering and maintenance staff, through to building managers and occupants
- influencing and specifying work to building services
- promoting wider sustainability – making the links with waste management and transport issues, for example
- energy surveys and audits
- surveying building service facilities and processes

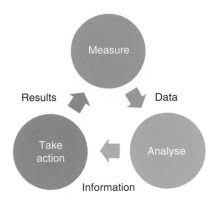

Figure 6.1 The energy management cycle

The energy management process

To be an energy manager, as mentioned above, you need the same kind of generic skills as a building services engineer. Additionally, it is beneficial to have a good knowledge of finance and financial systems. Strictly speaking, the energy manager should be concentrating on units of energy as a measurement, such as kilowatt hours (kWh), but the budget holders or financial controllers will want to translate these figures into budgetary systems of measurement. After all, although many of us will want to drive these processes to reduce environmental impacts, money is the bottom line imperative to most people and most processes within our society. So it's very important that the energy manager can speak both 'languages': finance and energy. Part of this process is presentation of information, which is absolutely key. Many energy managers I've known miss a trick, and don't present the data in a user-friendly, easily transportable way. There is a lot of similarity, as discussed in Chapter 4, between the measurement of finance and the measurement of energy. The spreadsheet is common to both, and is a vital tool for both management disciplines. Figure 6.1 shows the energy management cycle.

Data is everything

Without meaningful data, the process of energy management is half-baked or downright useless, so it's vital that good systems are put into place. Rather than reinventing the wheel, it makes sense to use good practice developed by energy managers over many years, so I recommend using standard practices and spreadsheets from, for example, the CIBSE Building Logbook Toolkit (CIBSE 2006), Action Energy Good Practice Guide GPG 348 (Carbon Trust

Customer Services
0800 056 7777
Mon-Fri 8am to 8pm
Sat 8.30am to 2pm

Emergency?
Electricity 0800 783 8838
24 hours a day, 7 days a week

You can make payments, supply meter reads or manage your account at:
www.edfenergy.com

Date	Bill for	Your account number
4 Nov 11	30 Sep 11 - 04 Nov 11	671

Page 2 of 3

Your statement in detail

Account activity since last statement
Your balance at your last bill was (29 Sep 11)	£129.24 cr
Payment 27 October 2011 - Thank you	£50.00 cr
Account balance before this statement	**£179.24 cr**

Electricity charges 30 Sep 11 - 04 Nov 11
Electricity meter number: 20f
Tariff: Online S@ver v7 Electricity (02 Feb 11 to 31 Dec 11)

	previous	latest	units	kWh split		price	total
30 Sep 11 to 04 Nov 11	18609 ●	19028 ●	419	419	at	9.36p	£39.22
Standing Charge				36 days at 22.88p			£8.24
Total electricity charges before VAT							**£47.46**

● = customer reading

Discounts and Surcharges
Direct Debit Discount	£2.85 cr
Total Discounts and Surcharges before VAT	**£2.85 cr**

VAT
VAT on £44.61 at 5%	£2.23
Total VAT	**£2.23**

Your total for this statement	**£132.40 cr**

Your energy usage
Your average daily electricity usage

At your last statement: N/A
This period last year: 13.68 kWh per day
This statement: 11.63 kWh per day

Your average has fallen or stayed the same

Billing detail

	Amount £

Fixed Charges
	£
Availability charge 250 @ £1.06/KVA	265.00
Standing charge	37.71
Settlement agency fee	2.50
Combined HH Data Charges	36.00
TOTAL FIXED CHARGES	**341.21**

Consumption Charges
	Price per Unit	Total Units	Amount £
Unit charge-Night rate	0.01858	19025	353.48
Unit charge-Day rate	0.03658	53533	1958.24
TOTAL CONSUMPTION CHARGES			**2311.72**
TOTAL ENERGY CHARGES			**2652.93**

Meter Reading Information
Supply Number	Current reading	Previous reading	Rate	Unit Type	Units	Constant
Supply point 001						
12000	478		Night rate	KWH	18745	
			Day rate	KWH	53187	
				MD	173	
				KWH	71932	
				KVARH	2025	1.0
Supply point 005	209226	207201				
12000	496		Night rate	KWH	280	
			Day rate	KWH	346	
				MD	0	

Account number	Period of supply	Date of invoice	Invoice
115	01 Sep - 30 Sep	02 Oct	0024

Figure 6.2 Comparison of a domestic and a commercial electricity bill

2003) and the CIBSE Guide F, Energy Efficiency in Buildings (CIBSE 2004); and other well-known organisations in the field can offer standard templates. Many of the calculations are already laid down and are common throughout the industry. Saying that, many of these templates can be adapted as needed, but are near enough universal. After all, a unit of energy is a unit of energy – it is the quality of the source data that matters. In my working life, I've seen so many extrapolated spreadsheets that have been guesstimates to fill in the gaps in the data. This is because bills have been missing or systems have not been put in place; this potentially leads to disaster because guessing the data leads to poor conclusions and unreliable judgements. With this in mind, there needs to be an emphasis on sourcing accurate and reliable data. There are a couple of methods that can be used to achieve this.

The first source of data is accurate data reading, either using online BEMS or collecting manually. There is nothing wrong with manual data, although for some reason it tends not to be trusted in the modern age, so long as it's taken at reliable, regular intervals – weekly or monthly. Even daily: the more data you collect the better.

The second source of data would be utility bills – historic bills if you're new to the post or are a consultant. However, there is a health warning with this, because the bills have to be accurate, whereas so many are based on estimates where readings have not been taken. These need to be approached with caution as the data can be worthless. Can you imagine a conventional company financial system based on guesses? Always question and check the source of the data and its accuracy.

Figure 6.2 shows examples of commercial and domestic electricity bills. The commercial bill shows considerably more data, for example including factors such as reactive power, kVA availability, and other parameters. The other problem that managers face is that, unfortunately, the way that the bills are laid out by different suppliers can often differ markedly, which can lead to confusion and makes data sometimes difficult to compare.

This issue also arises when tendering for contracts for utility supplies – it's hard to compare like with like. Companies present data in different ways, which makes it difficult to establish a level playing field. My advice is not to accept different companies' standard methodologies. When you're requesting tenders, ask the bidders to complete your own standard form, so that the data is all set out as you want it, in a like-for-like format. If you say that you will only accept tender bids on your own form, that solves the problem.

As with all data, the more often it is checked the more accurate it will be. Data on bills can be worthless, as already stated. A golden rule is to go back to the utility company for more data or a check on the data they hold. If possible, never accept an estimated bill from estimated meter readings.

The most important part about checking the bills and the data is to check the terms and conditions in place. For example, look at the availability charge and the maximum demand charges as specified in kVA supply. Looking at the illustrated bill, we can see significant savings that could be made at the stroke of a pen, without any great need for investigation. An example would be declared availability of supply. A building may have historically been paying an availability charge – it may be rated at 500 kVA, which might cost a thousand pounds. Do you ever use it or even need that level of availability? The building or the processes may have changed. You suddenly find you don't need 500 kVA, because you only ever use 200. By reducing the availability to 250 kVA, say, you've halved the bill for the kVA availability component. We once saved a large company £80,000 per year in this way. Similarly, on water bills, you pay a standing charge for the diameter of the water supply pipe. There might be a 100 mm main pipe and the size of meter that goes with this, which you pay for historic reasons and previous use. You may not need more than a 20 mm supply pipe now. Therefore, there's a chance to reduce the bill proportionately by 80% for the standing charge component of the pipe and meter supply.

To get to a scenario of making energy management popular with finance, and get low-hanging fruit, the best payback and quick wins, look at the terms and conditions first. In this way, you've saved money before you've even saved energy! You have made good savings and gained the confidence of the finance director.

The object of the whole exercise is to build a picture of the true energy and utility usage of the building, or of an industrial process. What you want to do is to get that reliable data and then start to break it down. By doing this, you can start the process of utilising monitoring and targeting (M&T). This is very similar, once again, to the financial process. M&T is a fundamental part of any effective energy management programme. Without this, the process is nearly useless. The aim is to present and review data to spot trends and anomalies. The key to a good energy management system is to get early warnings of waste and to quickly establish patterns of use for the building and its components, whether by department or by individual building services operations. This whole process is cyclic. Basically, you're continually collecting the data, exactly as a finance department would, looking for variances and established patterns of use, and utilising that information to target where you can save the most first. Using this as a continuous process to verify a savings process will deliver the goods. Ultimately, this is a win-win exercise where both the energy and financial managers get to control and accurately measure performance to develop more precise budgetary forecasts. Once a system has been set up on a basic level, it can then be developed to take account of other factors, such as measuring against degree-day data. A full explanation of degree days is given in Chapter 10.

You need to find ways of avoiding energy losses and you need a methodology to do this, so as part of data gathering you need a monitoring regime. You don't necessarily need to use a BEMS – you can carry out monitoring with a cheaper, cost-effective standalone system. Having said that, if you're a larger organisation with a package in place, then using energy management as part of that makes sense, and will probably be quite cost-effective as an add-on. Alternatively, if you're really on a budget, a good old-fashioned spreadsheet offers plenty of scope to create formulas as part of a management tool. There are some free-to-access basic examples available from the Carbon Trust – Energy Analyser Tool and user guide – these contain a number of spreadsheets that can handle data already recorded or provide blank templates for half hourly, daily, weekly or monthly energy data handling. This tool is also compatible with some commercial energy software (Carbon Trust 2010).

Graphs are also the energy manager's friend. Of course, nothing can be assembled in a graph without reliable data. With graphs, the old saying that a picture is worth a thousand words is apposite – you can create an instantaneous, easy-to-communicate picture of the situation that your finance manager will also probably appreciate. This graph can also aid in communication of the energy trends and performance for motivating the rest of the organisation to buy into the energy management and saving process.

One of the most useful graphs is the CUSUM graph (cumulative data graph) which shows a year-on-year picture of the performance of the building and its operations. This is what we want to achieve ultimately – the best way of measuring year-on-year performance, which can be easily communicated to business leaders. By using graphs, you can also immediately see anomalies and departures from good practice, allowing immediate action as part of an ongoing process. It's a diagnostic tool which can link in with a whole range of other issues such as planned preventative maintenance. All of this data and methodology can be fed into the whole process of the energy management system. In more recent times, we've seen the development of more formalised systems such as ISO 50001 and BS EN 16001. These are very much an extension of the quality and energy management systems – 9001 and 14001 respectively. These work on the same principle as plan-do-check-act. ISO 50001:2011, Energy Management Systems – Requirements with guidance for use, is a standard that aims to foster energy efficiency and promote best practice (ISO 2011). BS EN 16001, Energy Management, also aims to reduce costs and improve business performance.

Training, communication and education

The key for the energy manager is to take the data gathered and turn it into something you can communicate to a whole range of professionals in the organisation to gain resources and support, and to get buy-in from users of the building. How do we motivate and educate the users? This can be done in a number of ways, such as seminars, staff training, poster campaigns, incentivisation (e.g. reward systems) and inter-departmental league tables of energy. It should also be linked into financial performance systems, which can save thousands through the improved results. All of these can be linked together to make a very effective programme that achieves energy management savings. An extension of this, which would be very beneficial, is to also make it relevant to people in their domestic situation. If you show people how to save their own money in their own homes, then it will eventually become part of their culture to be energy efficient. However, we must be aware of a

Incentivising energy saving

A company had a big site with about 1000 people, including a large restaurant managed by a subcontracted catering company. This had been appointed through tenders to supply catering services, but the landlord and building operator was still paying the electricity bill. An audit of the kitchens revealed that because the catering company wasn't paying, zero energy management had been put in place. I recommended submeters, and after several months, energy use went down by 50%. This was because the manager of the catering company had to think about these costs for the first time. It is a custom in catering to come in and switch everything on, including the ovens and all the gas hobs. As soon as we start managing and measuring these costs, however, the incentive is there to save.

current trend in society where energy-efficient devices such as LED lights mean that people think that it's OK to leave them on all the time! To combat this, at work at least, it's sensible to relate power consumption to ongoing use and to ultimately tie in incentives to save energy to bonuses and other incentives which may motivate less committed staff. Small financial rewards can lead to big energy savings.

One of the things that we need to do is become amateur psychologists – how do we change people's habits? What's in it for them? Every means at our disposal must be used to achieve our objectives. If we can run workplace campaigns as part of any training in the building and make it as personal as possible so that people know how to save their own personal money, then that's a good motivator, and over many years of training I have found that it works. I once trained a group of maintenance engineers, and showed them how they could change their lamps at home. I explained how much beer money that would create in a year! At the tea break, one of supervisors came up to me and said: 'What have you done? Normally they're all talking about the football or what they watched on TV, now they're all talking about changing their lighting!' Being engineers, they also understood that the lamps would last ten times longer as well. Saving money was a clear motivator. This is an example of how we need to be imaginative and make things relevant to our audience.

The development of energy ratings

Energy ratings should be about measuring energy use in the total lifecycle of a whole range of energy-related plant, equipment and products, including all the things we use in buildings both commercially and domestically. As discussed in Chapter 1, I propose that the successful A–G rating system be adopted for everything energy related that we use in the construction industry. This can be coupled with minimum energy performance standards (MEPS), discussed later in this chapter.

This A–G system has been in existence since 1995, following the EU directive introducing the EU Energy Label, which has now become a widely recognised and respected guide for manufacturers and consumers alike. This current system covers a whole range of white goods, including fridges, freezers (Figure 6.3 shows an example of the improvement in energy performance of the same volume specification of a domestic fridge-freezer), washing machines and even ovens, and has been incredibly successful in educating the consumer to shop for more energy-efficient models. In fact, the system has been so successful that it is now rare to find any modern product that is below a C in the A–G rating. Because of this across-the-board improvement, to keep up with advances in energy-efficiency achievements, new ratings (A+ and A++) were introduced. More than this, the whole scheme may need to be re-evaluated to keep pace with this performance improvement. These new ratings were adopted from July 2011. As well as domestic white goods, the A to G rating is now applied to products including air conditioning units and pumps (Figure 6.4), as many contractors will have seen when they are installing equipment on site. In fact, A–G is applied now in a much wider range of industries.

Lighting also has energy labels, allowing comparisons that show, say, the benefits of tungsten lamps versus compact fluorescent units and LEDs (Figure 6.5).

Energy labelling and information is crucial for allowing the consumer and the specifier to look for the most energy-efficient and sustainable equipment.

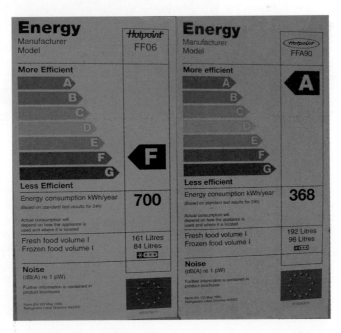

Figure 6.3 Fridge-freezer energy label

Figure 6.4 Energy label on a pump

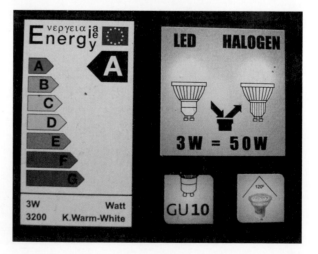

Figure 6.5 Energy label for LED lamp compared to halogen

I believe that these need to be encouraged for everything. In the building industry, the rating system has been introduced for all windows and doors that are fitted in the UK by the British Fenestration Rating Council (BFRC), who license an energy rating system (Figure 6.6).

Energy labels were also, via the US energy star efficiency rating, first introduced in 1992 for electrical items, and updated to now include televisions, which came into effect September 2011. Moving away from electrical goods, cars also now have an A–G rating based on their CO_2 emissions. This label also includes other information that enables consumers to compare the efficiencies and costs of different makes and models (Figure 6.7). The less

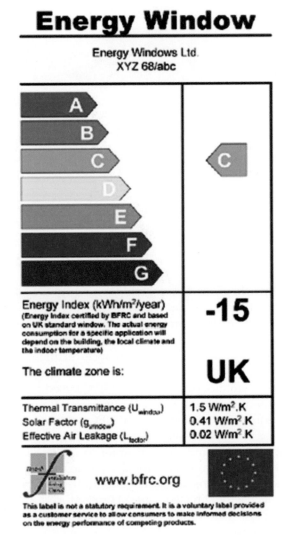

Figure 6.6 Window energy label

Fuel Economy

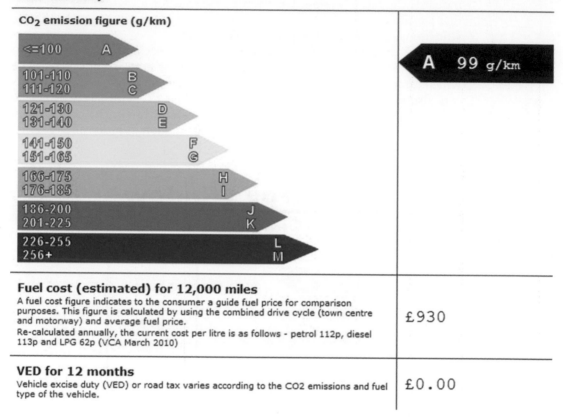

CO₂ emission figure (g/km)

<=100 A	A 99 g/km
101-110 B	
111-120 C	
121-130 D	
131-140 E	
141-150 F	
151-165 G	
166-175 H	
176-185 I	
186-200 J	
201-225 K	
226-255 L	
256+ M	

Fuel cost (estimated) for 12,000 miles
A fuel cost figure indicates to the consumer a guide fuel price for comparison purposes. This figure is calculated by using the combined drive cycle (town centre and motorway) and average fuel price.
Re-calculated annually, the current cost per litre is as follows - petrol 112p, diesel 113p and LPG 62p (VCA March 2010)

£930

VED for 12 months
Vehicle excise duty (VED) or road tax varies according to the CO2 emissions and fuel type of the vehicle.

£0.00

Figure 6.7 Car fuel economy label

emissions, the less vehicle excise duty is levied, to the level as can be seen of less than 100 g/km where no payment is required.

All of these previous examples are a very positive development for the wider building industry, and culminates with whole-build energy ratings, in the form of energy performance certificates (EPC – Figure 6.8) and display energy certificates (DEC). For example, there are now on-construction energy performance certificates and EPCs used for showing the energy performance when a building is sold or tenanted. For public sector buildings, it is a legal requirement to have yearly DECs.

This type of easily decipherable information enables designers, specifiers, contractors and the consumer to make a better judgement for selecting the most efficient choice for the particular application, At the other end of the scale in terms of clarity and transparency is the EU's Internal Market in Electricity Directive The EU introduced an electricity directive in July 2004 whereby electricity consumers had to be given information about where their electricity was sourced. This is only shown in the small print of electricity bills, in the electricity fuel mix disclosure.

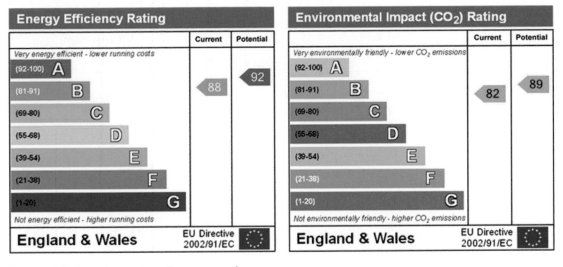

Figure 6.8 Example energy performance certificate

Table 6.1 Electricity fuel mix disclosure

E.ON	Electricity Source		
	E.ON energy solutions fuel mix (%)	E.ON UK overall totals (%)	UK average (%)
Coal	35.7	34.3	28.9
Gas	48.8	47	44.2
Nuclear	5.2	5	17.3
Renewable	6.6	10.2	7.9
Other	3.7	3.5	1.7

Source: EON Energy Electricity Bill (Domestic and Small Business)

This information is provided as charts or tables, and the number and types of electricity generation is listed as determined by individual EU member states. A UK example is displayed in Table 6.1.

This fuel mix disclosure allows consumers to differentiate between different electricity supply companies and allow them to make a choice when deciding to enter into a supply contract or to make a decision to switch supplier, whether based on price or seeking a 'greener source'. This can include a calculation of carbon dioxide emitted and the quantity of nuclear waste produced if nuclear power was used in the electricity's production. This information can be accessed on the utility supplier's website. An example is shown in Table 6.2.

I would point out that as with all statistics it's not always the full story. This gives emissions at the point of production in terms of the actual process of generation, but not the lifecycle or embodied energy throughout the whole process. It could be argued that both renewables and, in particular, nuclear will have embedded carbon that could be accounted for and disclosed at some

Table 6.2 Fuel mix including environmental impact

Fuel Supply	Ecotricity Apr 10–Mar 11	UK average (2010/11)
Natural Gas	24.00%	44.20%
Coal	17.50%	28.90%
Nuclear	2.60%	17.30%
Other	1.80%	1.70%
Renewable	54.10%	7.90%
Environmental Impact		
CO_2 emissions	266.6	450.3
Radioactive waste	0.00026	0.00173

point in the future to give an accurate representation. Simply saying that it is zero carbon in my view isn't good enough.

There are a number of other international and national energy ratings. For example, the US Department of Energy and Environmental Protection rating introduced the energy star rating, as mentioned before, which covered a range of products and equipment in the commercial and domestic sectors. In Australia, they have also adopted an energy star scheme based on the US version. In fact, New Zealand, Taiwan, Canada, Japan and the EU have also adopted the programme. Devices displaying the Energy Star logo, such as computing equipment and associated peripherals, kitchen appliances, buildings and other products, all generally use 20–30% less energy than required by US federal standards. The problem is, there are almost too many standards and schemes out there. If, at least on a national level, we agreed on an A–G system across the board, it would give us a standard approach to use when we measure and assess what is truly sustainable or at least working towards that goal. This, I believe, is a truly beneficial system. In Chapter 4, I looked at PER but this is limited to just one part of the energy cycle. It only rates one part of the lifecycle, measuring the conversion of the fuel into heat or power, not the sourcing or the ultimate disposal costs. We need one label for the whole lifecycle to give the complete information on the true environmental and energy costs, both economically and also with regard to the impact on resources and materials.

The EU introduced a directive on the eco-design of energy-using products in 2005. This has now been updated (Directive 2009/125/EC) which has been incorporated into UK legislation as the Eco-design for Energy-using Products Regulations 2007.

The directive aims to establish a framework to set mandatory environmental requirements for a wide range of energy-using and energy-related products sold in all of the 27 member states of Europe. More than 40 product groups are covered, including a range of domestic white goods, as well as TVs, lighting and boilers, which are collectively responsible for around 40% of all the EU's, greenhouse gas emissions. In 2009 the Directive extended its scope to energy-related products including windows, insulation materials and some water-using plant and equipment products. This was an important extension as it paves the way for more devices and can hopefully be integrated into a wide-reaching and transparent rating system to give a positive lead in making the use of lower energy rated products the norm for the future.

From a building services point of view, we use a significant amount of electricity for electric motors. In terms of the electricity generated worldwide, it is estimated that 45 per cent is used by electric motors across industry and the built environment (Waide et al. 2011). In fact, this accounts for all electricity produced in every second power plant, and therefore motor efficiency is key. In June 2011, a new EU directive was introduced: the Energy Using Product (EuP) motor directive which is a set of strict new standards for motor efficiency The international efficiency standard for motors was introduced to look at efficiency for low-voltage motors across Europe. A considerable number of other products and equipment are either in draft regulation or in implementation phase (BIS 2011). This can be checked at the BIS government website.

From a wider building perspective, there are whole standards not just on energy certification but which take a more generic approach. These go beyond conventional building structures and services to also take into account waste stream management, transport and other related corporate social responsibility issues. Examples include the Building Research Established Environmental Assessment Method (BREEAM), which originates in the UK (more details in Chapter 5) and the Leadership in Energy and Environmental Design (LEED) standard, which is the US version. Both these systems are utilised as third-party certification programmes to accredit buildings, but they don't look at or bring in energy labelling. They are about building structure and services, but they don't cover all of the equipment that contributes to major energy use in our buildings.

I believe that a good integration of energy labelling based on an A–G rating, combined with a structured methodology such as the Eco-design Directive, will make it possible to change and stimulate the market. Manufacturers of energy-using products and equipment will, from the design stage through to the manufacturing process, find innovative ways to reduce the energy consumption and associated negative environmental impacts, throughout their lifecycle. This will include materials sourcing, water use, energy in production and transportation, the reduction of any polluting emissions and the move to lower-carbon methods of use. Additionally, the end-of-life issues of waste and recyclability must also be a priority to truly make the market and consumers think about the wider issues of the longevity and impact of resource use, in their day-to-day lives and the buildings they occupy and work in. Ultimately, a combination of standards, legislation, market reaction and development will dictate the impact of this combination as we move towards the low carbon future; but until this becomes well and truly part of the culture, and is enforced, the overriding criteria will still be price and costs. I maintain that the market will react very positively because of the clear relationship between cost reduction and the efficiency of the use of resources. This will become the main issue and this will relate directly to the theme of energy hierarchy, which I highlight as a main issue throughout this book.

References

BIS (2011) Department for Business, Innovation & Skills http://www.bis.gov.uk/policies/business-sectors/environmental-and-product-regulations/environmental-regulations/eco-design-of-energy-related-products/eco-design-directive-regulations-and-studies (accessed 15.8.2012)

Carbon Trust (2003) *Action Energy Good Practice Guide* GPG 348 Access via: www.carbontrust.co.uk

Carbon Trust (2010) Energy Analyser Tool (V2.7) Available from: http://www.carbontrust.co.uk/cut-carbon-reduce-costs/calculate/energy-metering-monitoring/energy-analyser/pages/energy-analyser-tool.aspx#download (accessed 15.8.2012)

CIBSE (2004) *CIBSE Guide F, Energy efficiency in buildings*; access via: www.cibseknowledgeportal.co.uk

CIBSE (2006) *Building logbook toolkit and CD ROM templates*. Access via: www.cibseknowledgeportal.co.uk

HMRC (2010) *HM Treasury Carbon price floor, support and certainty for low-carbon Investment – Consultation*

ISO (2011) International Organization for Standardization Geneva, ISO 50001 guide, http://www.iso.org/iso/iso_50001_energy.pdf (accessed 15.8.2012)

Waide P, Conrad U. Brunner, Martin Jakob (2011) *Energy efficiency policy opportunities for electric motor-driven systems*, IEA Energy Series, Paris, France

7 Water – a forgotten issue

All the water on the planet has been here since the formation and subsequent cooling of the earth. The planet captured the water in its formation, and the water we use is the same water that was used by the dinosaurs and human-kind throughout the ages. Nature has her own ways of processing water, through rain, natural filtration and evaporation, and has recycled it thus for millions of years. We are using the same processes, but we use a lot of energy to accelerate the water cycle and production to keep pace with our needs and activities. The big difference is that we are using large amounts of carbon and energy for pumping and processing, which has wider effects and obviously costs us money.

Water and energy inexorably linked

It is evident from all that is written on the subject of sustainability that the changes needed to reduce our collective carbon emissions and extravagant energy use will need to be very significant. If the ambitious government targets to reduce our emissions by 80 per cent by 2050 are to be reached, then all of us are going to have to change our habits and accept that we will have to use less energy and be a lot more efficient. Consider our use of water. Many people forget just how much carbon and energy are used in its processing, transporting, pumping and ultimately taking away waste for discharge or water treatment. The UK water industry uses approximately 3% of the total national electricity consumption, or 4400 GWh annually with costs of around £200 million (Reynolds 2010). It is estimated that 70% of the energy used by a water company is for pumping water and sewage including aeration (CST 2009).

It's not just the actual water use; we also need to save the energy required to heat it. Water use reduction measures need to be coupled with ensuring that water tanks and pipes are well insulated and time clocks and thermostats are used effectively. This all has to be taken into account in the design and commissioning phase of all projects.

Water has not been generally regarded as important within the sector, and has become something of a Cinderella service. However, it was also been referred to as the 'new carbon', by Professor John Swaffield in his CIBSE Presidential address in 2008 (Swaffield 2008). All around the world, water is becoming a major resource issue. This is not just because of climate change, but also because

Delivering Sustainable Buildings: an industry insider's view, First Edition. Mike Malina.
© 2013 Mike Malina. Published 2013 by Blackwell Publishing Ltd.

of changing population distribution and growth. It is used intensively in farming and industry, and scarcity is having a bigger and bigger impact. Scarcity is one problem, but some places have too much water, while others get too little. This is true even within the UK. At the time of writing, major weather events have highlighted this: in the UK the north has too much water and parts of the midlands, south and east of England are officially in drought. Some parts of Australia haven't seen rain for five years, whereas many parts of Queensland have been under water in floods covering an area the size of France and Germany combined. Creating structures for moving water is a major economic and engineering challenge. Consequently, the infrastructure costs are enormous.

When it comes to water supply for individual buildings, from a mechanical and electrical point of view, the need to plan for it often gets neglected. Within the construction process, you have got to make provision for the supply and removal of water, but somehow it never seems to be high on the agenda; it should be.

The fact is that in the developed world, all the water that comes into a building has gone through a lengthy cycle of filtering, chlorination and pumping during the water treatment process. This means that water has a lot of embodied energy and carbon. Within a domestic situation, most of the water that goes in and out of a house is charged on a rateable value basis, linked to the property and not on a meter measuring the usage. Only recently have new houses been required to have a water meter. In the commercial and business sector, water meters were always fitted and generally compulsory. One would think that metering water would be an aid to saving it, but we are in the same situation as with all the other metering such as electricity and gas: even with the utility meters, people don't actively monitor and change their habits to save. Consumption is out of sight and out of mind. More so for many water meters which are often buried with difficult access and not always clear to take readings as in Figure 7.1.

Figure 7.1 A water meter; often out of sight, out of mind

Water meters are often quite literally out of sight, usually under a cover in a hole in the road, and is often underwater. Therefore, the average person never looks at the meter. Only the water company reads it, and often not for some considerable time resulting in many estimated readings. In the domestic arena, most people pay by direct debit and don't check their bills. They won't realise that you pay the same for water coming in as water going out, as water companies assume that 98% of the water that comes into a building goes out again.

It would be far better if water meters were more accessible and this tends to be the case in the commercial and business area (Figure 7.2). Also less common but highly recommended in larger premises is the fitting of in-line water meters (Figure 7.3), which will enable more accurate measurement and

Figure 7.2 Water meter located in an industrial unit a lot more accessible

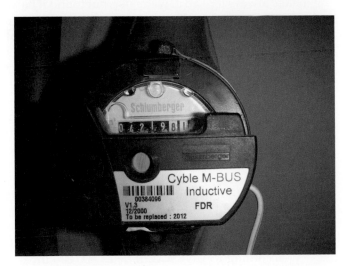

Figure 7.3 An inline water meter: can be used on the main supply in other areas in the building

> ## ⊘ How we work out your bill
>
> | Amount brought forward to this bill | £46.59 |
> |---|---|
>
> **Water you have used**
> Your meter readings have been estimated
> Meter number: 05A20
> Meter reading 22 Feb 11 : 6707   22 Mar 11 : 6719
>
> | Total cubic metres: 12 at 131.03 pence per cubic metre | £15.72 |
> |---|---|
> | **Total including VAT at 0.0%** | **£15.72** |
>
> **Standing charge**
> There is a standing charge for your water supply,
> based on your meter size:
> The charge is for the period from 01 Mar 11 to 31 Mar 11
>
> | Total including VAT at 0.0% | £3.35 |
> |---|---|
> | **Total for this bill** | **£19.07** |
> | **Amount due** | **£65.66** |

Figure 7.4 Sample taken from a metered water bill

management of water use to different parts of the premises, processes and activities.

Figure 7.4 shows a snapshot of a sample water bill. This shows a water meter is present and the measured use in cubic metres (1000 litres).

In the 26 years I've been carrying out building energy surveys and audits, I have always looked at water use, but found it hard to make the issues register as important with clients. Now that water is rising up the agenda and the changes brought in by Part G (Water efficiency) of the Building Regulations will make more of an impact and sharpen the focus on this vital resource. Water has also been propelled up the agenda with significant parts of the UK facing drought and potential stress on water supplies.

Some amazing statistics have emerged in the course of the debate over water, for example:

- The South East of England has less water available per person than the Sudan and less than most European countries.
- The average Briton drinks three litres of water per day and uses 145 litres for washing flushing and cooking. In reality, the average Briton actually consumes 3400 litres per day because of the amount of water used in the cultivation and processing of our food, before it gets to our table. About 70 per cent of all water used by humans goes into food production. And, according to UNESCO, it takes 400,000 litres of water input to manufacture a single car (Chapagain 2004).

Water bills are set to rise significantly because of the need to improve the infrastructure, and also because the water industry uses so much electricity to process water, so the rises in electricity prices will affect water costs.

Making the connections between water use, energy, carbon and cost has not been high on people's agenda to date. This is set to change. Along with the other elements of building services that we are examining, redesigning and reappraising water is set to become a big issue. For the first time in 2010, as in Part G of the Building Regulations, water efficiency calculators became compulsory for new dwellings. There are no equivalent standards in the UK for commercial buildings, but this will have to change in future regulations and at the time of writing is being proposed.

Just as the A–G rating for energy as mentioned in previous chapters is well established, the idea of a system to measure and label water usage is a welcome development. The government department DEFRA has encouraged the Bathroom Manufacturer's Association (BMA) to devise a water efficiency rating system that is simple and easy to understand. This covers a range of water-using products from sanitary ware, showers, baths, urinal controls and replacement WC flushing devices. Grey water recycling units are also included. Figures 7.5 and 7.6 show example labels for water efficiency ratings on showers and WC suites. More products are also being labelled as water efficient as development takes place and this will hopefully stimulate the

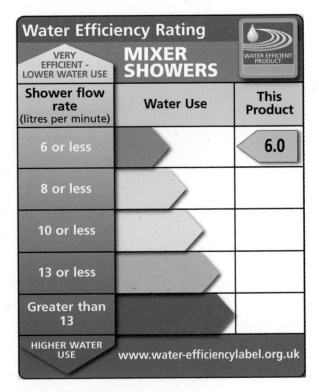

Figure 7.5 Water efficiency shower label

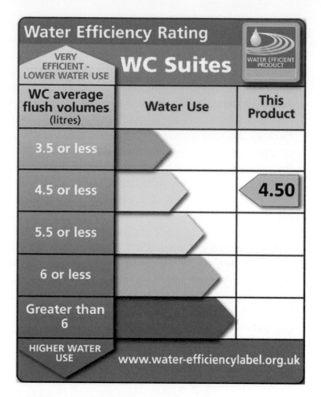

Figure 7.6 Water efficiency WC suites label

Figure 7.7 Water efficiency WC suites label

market and product manufacturers to accelerate more efficiency in the sector and increase consumer awareness (see Figure 7.7).

The industry has also produced a water calculator which can be used with Part G of the Building Regulations. The calculator can be accessed at: http://www.thewatercalculator.org.uk

In the UK, 52% of water is used domestically, of which a third is used to flush toilets. In Australia, because of water scarcity, they can now use dual flush toilets which flush using either 4 or 2 litres, whereas past and traditional systems in the UK use up to 7 litres. A third of water being used for one purpose is significant enough, but in some commercial buildings with urinals, continual and unregulated flushing can account for up to 60% of water usage. These systems need water-saving devices fitted – there are now many products on the market that can work on presence detection to effectively reduce water use with very significant financial savings and payback. Otherwise, urinals are expensive and wasteful in conventional water usage. Figure 7.8 shows an automated water-saving device.

Many of these mechanically based devices, such as the one in Figure 7.8 and fitted from the 1970s to the 1990s saved water by reacting to water pressure changes. If someone turned the tap of the hand basin on, the urinal would flush if these were set up correctly. Sometimes very subtle variations in water pressure or from a neighbouring system would also activate these devices. Unfortunately, many of these systems fitted in hard water areas weren't maintained correctly, so when scale built up, they would stick either closed or open, meaning that the urinals were either never flushed (certainly saving more water!) but gave rise to different problems, or else they were flushing all the time, as if the cisterns had never had a water saving device fitted. More recently, many devices have been linked to infrared presence detection which could also control lighting and extractor fans (Figure 7.9). These still need to be maintained with a good water treatment regime (see Chapters 10 and 12).

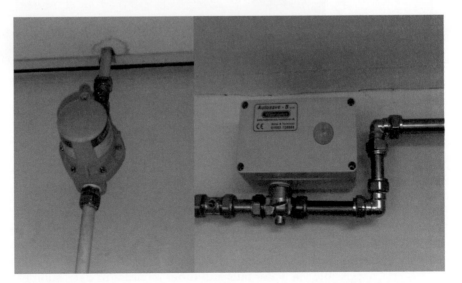

Figure 7.8 Mechanical pressure activated (left) water saver and an electronic presence detection device (right)

Figure 7.9 Several urinals controled by a presence detector (as indicated) linked to an automated valve

Management of water in building services

The management of water in building services requires a very similar process to the management of energy, and it works on the same principles. If we don't measure consumption, we can't manage it. Therefore, the first priority for the facilities manager or building services engineer should be to find out how much water a building is using and to obtain as much historical data as possible via water bills and other records. To date, submetering has not been installed in most buildings, so at some point it might be useful to consider fitting water submeters to strategic areas, particularly any high input or high consumption areas, such as cooling towers, swimming pools, process plant and catering use. Just as with energy use measurement, you need to break down usage by areas and activity to fully understand it. In the same way as for energy, therefore, smart metering is being developed for water usage, although it has had less impact than smart energy metering thus far. Ultimately, however, it's the same technology and it can be used very effectively if it is linked to a BEMS (as in Figure 7.3, and covered in Chapter 10 on building controls).

As with monitoring energy use, a pattern will start to emerge. This is the only effective way to start a proper programme of water management. Essentially, it's the same as any management process – you measure, manage, control and reduce. The only way to see if usage is high is to compare data with other similar buildings, so it's useful to benchmark. There is information on how to do this which can be accessed by searching for 'benchmarking' on the Envirowise website at: www.envirowise.org.uk

Unfortunately, despite the similarities in approach between energy and water, there are currently two organisations responsible for promoting government policy and initiatives on energy and water respectively. The Carbon Trust deals with energy while Envirowise is responsible for water. Sometimes, therefore, the crossover is missed. This seems to me to be a major error, and I believe that the two should be brought together. After all, the water process cycle is a major energy user, and has a large carbon footprint. The two should not be separated.

In any case, benchmarks enable you to build a picture in order to gauge usage. There are various ways of doing this, based on the type of activity taking place in the building and the number of people using it. A good example of benchmarking is the comparison of schools. During the resource audits that I've undertaken, I have found almost identical schools, in terms of composition, building structure and numbers of staff and students, where one is using around three times the amount of water of the other!

This could be down to a number of factors. It can literally be the difference between a school caretaker actively managing (or not) the water by simply turning the taps off, and another establishment actively fitting water controls. Most schools now have percussion taps or press taps, which shut off automatically. These significantly reduce water use. Others will have presence detection, as mentioned earlier in the discussion on urinal controls.

Sometimes I've found that water is used most in the school grounds, where you might find someone watering the football pitch in the middle of the day! Of course, this should actually be done on a timer out of hours, so it's done at a time of day when there's least evaporation. Some schools also use water for plants and shrubs. Therefore, planting drought-resistant plants could make a major difference to water use. Utilising rainwater capture, discussed later in this chapter, is also another important issue to consider. Even then it is important to use this precious resource wisely. A simple clockwork water timer fitted to the tap feeding the hose provides effective water management as shown in Figure 7.10.

Figure 7.10 A simple 'clockwork' water timer

These are the sorts of issues that need to be examined across any building. In the average office block, water use is almost entirely concerned with the toilets and the catering, whereas in manufacturing, it will be used for a whole range of processes depending on the manufacturing activity. Most people forget the amount of energy that's used in heating hot water, but this also needs to be counted as part of any water management saving strategy and integrated into the overall energy management cycle. One of the key and easiest methods is to reduce the flow rate in non-critical building services. Basically, if you reduce the flow rate, you restrict the amount of water that can flow. People can use water for the same length of time, but they'll be using less. This has to be managed very effectively, sometime by trial and error, because if it's restricted too much, then you may lose the support of building occupants. They may see it as too much of an inconvenience. This kind of intervention needs to be done either gradually or in conjunction with education for the staff and users of the building, or ideally both. This should be fundamental to the education of building users which I mentioned in the Introduction and will outline in more detail in Chapter 14, which deals with changing behaviours.

Grey water and rainwater harvesting

The current trend is often to look at innovative or novel ways of managing water such as grey water and rainwater harvesting. The latter captures rainfall from roofs using the guttering and downpipes to channel the water for collection in a water storage tank. Grey water collection involves taking non-sewage wastewater, such as water from hand-washing or bathing, and diverting the flow away from the normal route for disposal via the foul water drain. This has to be treated very differently.

Both these interventions sound like good ideas, but there can be significant costs, which hardly stack up on economic grounds. They are more likely to be used or deployed as good practice, environmental measures. In reality these projects in certain circumstances can be very similar to what I refer to as 'green bling'; a good example of this being photovoltaic panels, often employed as an option before considering basic and far more effective measures both on cost payback and energy/carbon reduction effectiveness. People like the idea of being green, but these projects can be a very big distraction from reducing water use in the first place – the approach at the top of the energy hierarchy which we discussed in the Introduction. Because of the similarities between water and energy, this hierarchy could be a water hierarchy just as much as it is relevant for energy. The key is to reduce the amount of usage in the first place and not to use other methods to justify current use.

It could be argued that the embodied energy and cost may not quite make it a viable option, although if everything else has been covered on the water and energy hierarchy then this may be an option to consider. It certainly focuses as a talking point and visual education and as a reminder of the importance of water as a vital resource.

I have experimented with rainwater capture, as part of my attempt to practice what I preach (more in Chapter 15).

In the right circumstances, grey water and rainwater harvesting may be good applications. Still, before jumping to try them we need to consider several factors. To begin with, we need to take into account whether we're looking at a rural or an urban location, and the physical nature of the building or site. This is because these initiatives will involve new infrastructure and storage. They also involve extra pumps, which of course require energy, which is something to be taken into account when aiming to be green. Additional infrastructure also means more embodied energy and carbon, which must be considered for its impact and whole-lifecycle costs and resource implications.

It also needs to be remembered that although rainwater is clean water when falling, at least outside of the cities, as it lands it picks up contamination on roofs and in gutters, whether from leaf detritus or bird waste. There are, therefore, important health implications, which is why we chlorinate and treat water in the first place. The same applies to grey water, insofar as it is, in effect, contaminated water. Therefore, to use either rainwater or grey water may require additional filtering and possibly additional water treatments,

Figure 7.11 School rainwater harvesting project

depending on what the water is to be used for. Sometimes, this isn't thought through in relation to the wider issues of water use. This means that it can become a far more expensive way of using water. There are also the additional considerations of increased maintenance issues of roof and gutter cleaning. This all needs to be examined and considered in any new build or refurbishment project when put forward as an idea at the design stage.

To use rainwater harvesting effectively, it's often better applied to uses outside of the built environment, such as the irrigation of gardens and school grounds. A good practical example would be water butts (Figure 7.11). Water, if even mildly contaminated, will easily become stagnant and cause problems.

A mesh or screened filter on the feed and top of water butts will help prevent mosquitoes laying their larvae on the top of the water – otherwise, this is a good example of another health hazard or nuisance that is often missed.

There are some systems of rainwater and grey water harvesting that are not only using extra pumping, but because of the perceived health implications are using ultraviolet lamps to kill any microbiological material and thus clean the water, which of course is using energy that wouldn't otherwise have been used. In my view, this does rather defeat the object of the exercise. This needs to be weighed up as part of the embodied energy and lifecycle issues. However, one day we may need to use additional energy onsite to clean our own water, not because it saves energy per se, but because water will become such a valuable commodity. This is, of course, because of the impact of climate change, and also because we are using so much of this precious resource.

Water and pipework infrastructure

The thing to consider is that all the water we use in the western world is, by and large, pumped on a one-pipe system. This means that *all* the water we use has been treated and purified. However, we only drink less than 1% of this. Ideally, therefore, if it were possible to have a two-pipe system, that would be a much better way of doing things. The highly processed water, with all the embodied energy, is really only needed for drinking. For everything else, we should be using less-processed water. If we could start the whole system from scratch, knowing what we know now, we would do that, but we have an existing infrastructure, which in some cases is well over 50 years old.

Some new developments will be using more grey water, which means that all the health considerations I have just mentioned must be taken into consideration. We must be aware that, by installing extra infrastructure and tanks, filtration etc. for grey water, we are locking up yet more embodied energy into the process. The big question is: can this be offset by water savings in the infrastructure?

These are decisions that will have to be made ultimately by the client, who is paying, and in some cases it would be used more for educational value than for real-life energy and carbon savings. In demonstration projects, the value is in making people think about the whole issue of water resource usage and its conservation.

Never forget that you could end up using way more energy in capturing the rain. If you carry on using the water, yes you might be wasting it, but from an embodied energy view you may be using less than extra infrastructure. It's a difficult call because, as we have already stated, water doesn't just flow out of the tap – it has an awful lot of embodied energy. One option might be to reduce energy in water harvesting by using gravity, which would entail raising the rainwater capture storage so that pumping becomes unnecessary.

This whole topic also relates to the issues covered in Chapter 2, on planning ahead. Planning considers the issues of building use and appearance, but must also consider issues of infrastructure, taking into account the balance between energy flow and infrastructure costs. As I mentioned earlier, water will become an increasingly valuable commodity because of the fact that it will become scarcer. People will have to accept that they will have to pay a lot more for it, and that may be the only way to incentivise people to save it. This could, in the future, lead to a whole range of changes in society, even to the extent of the type of food we eat, because of the amount of water locked up in its development and processing. The use of water is cumulative. For example, 1000 litres of water is needed to produce 1 kg of wheat, but 1 kg of beef needs about 15 times as much water (Waterfootprint 2012).

Pressures on infrastructure, coupled with climate change, will likely lead to greater water scarcity and rationing, which has already happened on several occasions through the last few decades. In some parts of the world, this is

Figure 7.12 Irrigating agricultural land will be under more pressure

Figure 7.13 East Anglian irrigation ditches looking very dry

already a very big issue, and those countries that have the resources are dealing with this through an increase in the use of desalination plants. These, however, require an incredible amount of energy. It has even been suggested for the UK, but the question is: where is the energy going to come from? Surely it is better to invest as much as possible into managing the resources effectively and conserving the water, as well as modernising to create an effective water distribution infrastructure.

At the time of writing in March 2012, this issue will be a pressing issue for the people in the East Midlands, South-East and East of England, as we are ending a winter with the driest period for 40 years (Guardian 2012). Unless my home region of Suffolk gets significant rainfall in the near future, then water restrictions are very likely. This is a significant agricultural region, especially encompassing the Fens and Brecks of East Anglia. The Environment Agency is likely to impose severe water restrictions and perhaps rationing (Farmers Weekly 2012). This will have a big implication on the irrigation of crops and on the landscape (Figure 7.12). Irrigation ditches look very dry (Figure 7.13) and the lack of available water for crops will have a big knock-on effect to the local economy as well as a direct impact on crop yields and food prices for all of us. As a developed nation with significant resources, we will be able to get by, but it does give an indication of one of the big issues for the future and a slight idea of what many parts of the world are going through already. This is covered further in Chapter 16, where we look at sharing our technology and expertise with the developing world.

This highlights the importance of making the connections of water use and sustainability, as well as the links with energy, resource use and climate change.

References

Chapagain AK, Hoekstra AY (2004) 'Water footprints of nations, Volume 1: main report', Value of Water Research Series no. 16, UNESCO-IHE

CST (2009) *Improving innovation in the water industry: 21st century challenges and opportunities*, A report by the Council for Science & Technology

Farmers Weekly. 'Water restrictions likely after driest winter', 15 February 2012

The Guardian, 'Half of UK households could face water restrictions by April', 15 February 2012

Reynolds LK, Bunn S *Improving energy efficiency of pumping systems through real-time scheduling systems; Integrating Water Systems*, Boxall & Maksimovic, Taylor & Francis Group, 2010 London

Swaffield J (2008) 'Living with the albatross', CIBSE Presidential Address 2008

Waterfootprint http://www.waterfootprint.org (accessed 16.8.2012)

8 Putting it together – the contractor's role

The role of the contractor in the construction process is clearly crucial. After all, the contractor puts everything together. Contractors are also vital for the efficient running and maintenance of a building, as well as for repair and refurbishment projects.

There are many able and skilled contractors in the marketplace, but too often they don't get the opportunity to input fully into the work that they are charged with creating. In essence, the process falls apart because the contractor, more often than not, is brought in too late to impact upon the design phase. This disadvantages the entire construction process.

Giving contractors room to work

Most of the time, contracting is still done by competitive tender, whether the contract is for mechanical or electrical work. A tender document containing the specification, often compiled by a consulting practice, goes out to a number of contractors. The winner is usually the bidder who comes in with the lowest price. This seemed to be the logical way of working for many years. It has, however, increasingly come into question. The system can let us down. The biggest problem is that when the submissions from the contractors come back to the consultant charged with evaluating the bids, it's very hard to know if they are truly comparing like with like. The contractors are replying to the same specification, but elements can be missed, and different methods for delivering the specification might be planned. Also, the specification just might not be tight enough to ensure similar interpretations. Therefore, it's very difficult for the evaluator to say with any certainty that all the bidders are offering exactly the same service, methodology and product.

One example might be a contract for design and build. It may well have a whole series of clauses, on desired operational parameters and outcomes, but very little on the actual technology to be employed. It might say merely that the client wants the air handling system to achieve a desired temperature range. In cases like these, it's clearly impossible to evaluate like for like. Potentially, the same specification could be answered with completely different designs and systems. If a client wants to achieve a sustainable

Delivering Sustainable Buildings: an industry insider's view, First Edition. Mike Malina.
© 2013 Mike Malina. Published 2013 by Blackwell Publishing Ltd.

building, however, it's crucial for building services managers, consultants and architects to be exact in the specification. Too often, clauses are just cut and pasted from the previous contract! In the end, this puts everyone at a disadvantage. The end product costs a lot more, things get missed and changed, and then there are the inevitable variations to the programme. Variations can cost serious money.

So what's the answer? Competitive tendering needs to be done on the basis of a complete design specification, not just with a simple design and build brief, which enables corners to be cut to reduce short-term costs. From then on, it's about following a whole series of steps, to maximise the efficiency of the construction process, minimise misunderstandings during the project and make sure that all elements of the specification are coordinated and agreed right from the outset.

I consider the development of BIM as covered in the introduction to this book, and the process of 'soft landings' as developed by BSRIA, a great step forward in helping to achieve a coordinated project and process covering most of the buildings lifecycle.

This general process was first mentioned as part of the FIT Buildings network project (FIT 1999) part funded as a research project by the UK governments, Department of Trade and Industry (DTI) in 1999, under the DTI Partners in Innovation scheme (PII). I was a member of this project team and saw the value of this whole concept and approach from the earliest stages. I was pleased to participate and provide several early case studies on significant refurbishment projects I was working on, which encompassed all the principles of the process as described in Figure 8.1. It was also at this time that we also first started to look at the concept of continuous commissioning (see Chapter 10).

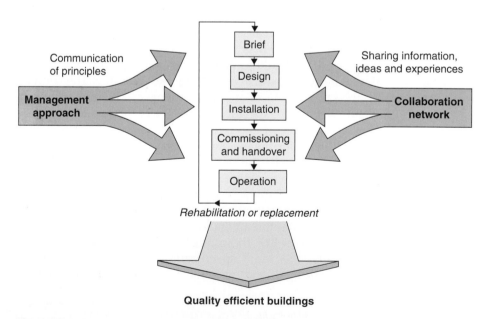

Figure 8.1 Process to achieve quality efficient buildings

Now more than a decade on from this pioneering set of projects, I am pleased to see many of these early ideas being developed significantly from the original series of partnerships and test case studies, which laid firm foundations for what is now becoming mainstream throughout the building services industry and further afield.

This project was developed shortly after the Egan Report – Rethinking Construction: The Report of the Construction Task Force (Egan 1998), which was very much the progressive thinking of the time and sought to improve performance through eliminating waste from the construction process. Part of this was the focus on seeking to foster several drivers of change including the vital process of cooperation from all sections of the construction industry. We certainly still have some way to go, but history has a tendency to repeat itself and we need to learn from it. This early project concentrated on the concepts of an integrated project process, very much following the stages as set out in Figure 8.1. The added dimension that developed was a very early and pioneering forerunner of what would now be described as sustainable or a low energy/low carbon approach to engineering. At the time we called it the E-Co management approach (FBnet 2000).

For me it's the getting-things-done part of the process that makes or breaks the project. Beyond the vital brief and design – and this has to be right from the start – is the role of the specialist contractor, which is a fundamental part of the project. The smooth process of installation and the interrelationship and cooperative working on a project are what makes it work well. All through these stages is the process and development for creating a good regime for commissioning and handover through to the successful operation of the building. This is where the original concept came from, with the development of continuous commissioning, communicated by a sound management approach and the collaborative network that it created. I have seen this work and experienced the success of projects through this methodology. I have also seen what happens when this isn't carried out and it's what can give the industry a bad name and drag down the vast majority of committed professionals, who want to achieve the quality efficient buildings that we need and must achieve on the road to the emerging low carbon society of the future.

The concept and process of continuous commissioning is about getting best value from our buildings by:

- seeking opinions from building occupants and users
- responding with appropriate actions and making sure that the building services are maintained and optimally adjusted for best performance and to achieve a good indoor climate in which the occupants can work and feel comfortable
- measuring the results of the occupants' responses together with providing continuous improvement and fine tuning through continuous commissioning.

This, I'm glad to say, has been continued as a concept in the BSRIA soft landings process.

The soft landings process

The soft landings process was developed by the Building Services Research and Information Association (BSRIA 2009) to smooth the entire construction process and to mitigate the problems and discrepancies that arise. Soft landings requires designers and constructors to engage in dialogue throughout the process and to stay involved with buildings beyond practical completion. This assists the client during the first months of operation and further into the future, to help commission to optimum performance and to eliminate any errors in the building services systems. This process is also important to ensure that the occupiers understand how to control and best use their buildings.

Soft landings documentation covers the duties of the construction team during handover and for the first three years of occupation in the following stages:

Stage 1: Inception and briefing

Compared to the conventional building approach, the process allows more time for constructive dialogue between the designer, main contractor and client. Without soft landings or a similar process, the convention would be that the architect or designer would draw up the plans and elevations, without systematically integrating the details of the building services elements. The consulting engineer who designs these elements may meet the designer once, if that, and may not even have used the same package as the designer to draw their portion of the plans. Also, when the mechanical and electrical (M&E) subcontractors are appointed, they traditionally have little contact with either the consulting engineer, the designer or the client. With soft landings, on the other hand, the project team sits around the table from day one. This involves the designer, main contractor, consulting engineer, subcontractor and client's representative. Everyone has to agree and buy into the entire process, and one person acts as the commissioning or programme manager. They draw up a formalised project plan and record minutes with action points from regular meetings. Typically, these would be held every Monday morning. By adopting this process, the team will hopefully eliminate any management problems. There will always be minor discrepancies in any project, but this approach will go a long way towards minimising them.

Stage 2: Design development and review

In the past, project teams have tended to have a fairly narrow focus – it's just about getting the job done. With soft landings, the team has to do rather more. For instance, when the M&E subcontractors are brought on board, a regular meeting process becomes part of the contract. The subcontractors have to commit to these meetings and sign up to the process. This reinforces the project and maps out each stage and everyone's role in the construction process. The whole idea is for the participants to look to the future, to visualise how the building will operate in practice. They have to look at energy use and general usability during the construction process. This approach dovetails in with energy performance, corporate social responsibility (CSR) and documentation such as building logbooks. These are now mandatory under Construction Design Management Regulations (CDM) and Part L. In short, it brings the entire project team together to review insights from comparable projects and to detail how the building will work from the point of view of the manager and individual user.

Stage 3: Pre-handover

In the conventional building process, this stage often simply doesn't exist! Contractors finish their work and leave site without a formal pre-handover stage. This stage allows the M&E contractors to go through what they've done, with the project manager or a nominated client representative. They will then have all the documentation and instructions in one place. This documentation will cover both the maintenance regime and the energy performance. Therefore, the future occupier can be trained on how everything should work. This is the resource intensive part of the process, and it's often not done because of the extra cost it entails. Developers are sadly often not interested in this stage, since they are not going to occupy the building themselves. However, they should consider that doing this would still add value for the prospective buyer or lessee. Surely, it would attract customers to offer them information on the smooth running of the building and, importantly, how to save money by operating it correctly and efficiently? It's slightly more expensive in the first instance to perform this exercise, but it will certainly add value. Therefore, if the architect and engineer can persuade the client to do this, then everybody wins.

Stage 4: Initial aftercare

This is what I call the continuing commissioning process. On a standard build, everyone is gone by this stage, and they will only return for snagging or dilapidations, if required. Using the soft landings methodology, on the other hand, the handover and initial stages of occupancy should run smoothly. People don't just run off to another job; they work through the issues. The client gets added value because right from the start, they know that everything will work as intended, that a maintenance regime will be put in place, and that the energy performance will be good. This will obviously minimise costs in the longer term. At the moment, on an average project, this is usually not done. People are just left with a set of manuals, and get half a day's training if they're lucky. The benefits of using the soft landings approach should be clear by contrast.

Stage 5: Years 1–3 extended aftercare and post-occupancy evaluations

This stage is about finishing the process and ensuring that any new projects will be easier to manage and meet the expectations between of the initial design and the reality of actual performance and operation. It is an extension of continuous commissioning, encompassing planned preventative maintenance and aftercare. This is developed further in Chapters 11 and 12, which deal with commissioning and maintenance respectively.

Buildings have to be maintained or performance drops off considerably. This process of aftercare is like maintaining car – you need to give it a regular service or you will get bad performance. In the same way, buildings need to be checked weekly or monthly, particularly with regard to their controls. There also needs to be continuous professional development (CPD) for the building operator regarding any changes – upgrades of the building management system, for instance. At the moment, after a BMS has been commissioned, people are often just left to operate it with very little training (see Chapter 10). Then, when someone leaves, there will be a new recruit with no knowledge of how to operate the particular system. Anyone new, clearly needs to be brought up to speed with the system. Therefore, this process of aftercare can be a much better approach to managing systems.

Towards proactive contracting

It is important for a contractor to be proactive, and not to wait until tenders arrive out of the blue. All contractors should try to pre-empt the market, to find out what's happening, and to seek out architects and building services consultants. In this way, they have a chance to establish themselves as experts to help in the design process. Otherwise, they will probably get no input at all. However, if they do maintain contact everybody wins, because there is potentially so much added value if they can input at an early stage. It can save clients, architects and consultants a lot of time and money. As experts in realising designs, they are uniquely placed to be able to neutralise a lot of problems in the first place. This is confirmed by looking at Figure A, design cost impact stages and influence, in the Introduction; this illustrates the benefits of early contractor involvement.

Unfortunately, many contractors tend to be reactive rather than proactive. One of the ways they might change this is to do a lot more work on market intelligence. It pays to attend trade body meetings. A good example of this kind of opportunity would be the events put on by the B&ES, as well as the Electrical Contractors Association (ECA). At these meetings, there is a lot of discussion of projects and technologies, and insider knowledge on what is happening in the industry. These networking opportunities are invaluable. Many contractors say that they don't have the time, but they could gain a lot of benefit from these meetings.

As well as attending meetings, contractors should be looking to put their own house in order. All of their processes should reflect their commitment to best practice. If their own systems follow good practices, the supporting documentation will dovetail in almost effortlessly when they bid for tenders and contracts. For many contractors, there are a lot of lessons to be learned from an exercise in changing their internal culture.

It's also important for a business to reflect its ethos in a tangible and meaningful way. For instance, from a sustainability point of view, how sustainable is a contractor's own premises? Every contractor should look at their premises as an opportunity to create a shop window of the benefits of a sustainable building. Good premises can work as an advert to show what can be done. Perhaps the saying is true, as one contractor replied when this was pointed out to him, that 'the cobbler is always the worst shod'. Perhaps contractors get too close to the situation. Perhaps they are always so busy with other people's buildings that they forget their own. But, in addition to creating a powerful advertisement for their business and their skills, if they get a really efficient building, one which looks good and operates well, they will also deliver savings to the bottom line of their organisation. Every pound saved on fuel bills goes straight to the bottom line. That's why energy efficiency is such a good practice and investment. Figure 8.2 gives an example of a simple but effective message to business by East Anglia based contractor, Pitkin & Ruddock.

A good example would be a contractor who installs heat pumps. It would make sense for them to heat their own offices via a heat pump, if practically possible. It shows that they believe in the technology that they are trying to sell and install. Also, from a cost point of view, they can get the technology at trade price. If they do have any downtime, it's also a great way of keeping employees busy, and because it is in constant use on the premises, it serves to reinforce the best operation and performance of the equipment, especially as the contractor's own staff will be users of the technology.

I like to term this whole process of proactive relationship building, networking, best practice processes and showcase premises as 'added-value contracting'. Contractors should be turning themselves into enhanced-value enterprises. That way they would be developing a much firmer base for their business. If contractors and contracting companies followed this process, it would deliver them significant benefits. It's an opportunity to develop a competitive edge in the new low carbon economy, and to enhance their business reputation. Those who engage with this opportunity will be known for going that extra mile. Construction and building services engineering is, in a sense, a small industry. News travels fast. A contractor is only ever as good as their last job, and reputation is everything.

A sustainable future for contractors

The biggest element of the whole developing sustainability market will be the refurbishment and retrofit segment of the industry. So often, too much emphasis is placed on new build. However, of the buildings that will have been constructed by 2050, upwards of 85% of them, have already been built. The

Figure 8.2 An example of a simple but effective message to business

industry looks to the prestigious new builds, but the real challenge is sustainable refurbishment. This is the market that will really make the biggest impact. The opportunity for retrofitting is huge, both in the commercial and the domestic arena. For example, the housing market is going through a significant growth in refurbishment. The UK is full of old Victorian and Georgian houses, and the appetite for improving their performance is growing because people want to save energy. Old, inefficient buildings have got to be improved. What will drive this even further will be the implications of increasing fuel bills. Oil, gas and electricity prices will all rise over the next decade, making retrofitting more and more cost effective. This applies to schools and offices as well as houses.

Furthermore, legislation will require improvements to be made. The next set of building regulations could be set to say that if a homeowner builds a conservatory, they will have to spend a sum on consequential energy improvements. Therefore, it makes sense that if you are undertaking any project in the building, you improve the energy efficiency of its other parts. Consequential improvement is currently only compulsory for commercial buildings of over 1000 m^2. In these buildings alone, both the European Directive (Article 6 of the European Energy Performance of Buildings Directive (EPBD)) and the building regulations currently state that if work is being done, 10% of the total cost, over and above the sum spent on the project, has to be spent on consequential improvement of the building, to improve the existing building energy performance (CIBSE 2003).

The B&ES estimates that the market for refurbishment is worth £5 to £8 billion to the building services sector (HVCA 2011). Refurbishment offers fantastic opportunities for contractors and engineers. These opportunities are not just around business development – there is also amazing scope within these projects for skills upgrades and technological development. A new series of technologies and techniques will have to be developed to meet the needs of sustainable retrofitting. For example, single-skinned walls need lower-cost insulation techniques. This is just one of a great many challenges to create and apply new techniques. Ultimately, this is a long-term market opportunity, because it will take decades to bring all the existing building stock up to standard, both in the UK and beyond.

To achieve all this, contracting companies will have to invest heavily in training. Many people think of the sustainability market as a new industry, but it's actually an adaptive industry. All the basic skills and knowledge are already in place. What contractors have to do is to upgrade and adapt what are existing skills.

For renewable energy and lower carbon technologies to be encouraged and deployed, a lot of the stimulus for this market could come from the feed-in tariff and renewable heat incentive. This was subject to volatility, with government indecision, during the first half of 2012, but could still have the potential to help take this market forward.

These technologies, such as renewable and low carbon fuels, will become more cost effective as carbon fuel prices rise. By not using carbon fuels, therefore, you get a double benefit. Contractors need to become conversant with this. They need to be able to explain the FIT and the renewable heat incentive to end users and customers. This will then be a major stimulus for their market.

Market mechanisms, incentives and legislation are the combined market drivers for this adapted renewable and low carbon industry. The key to the whole adaptation for the construction industry will be to make sure that standards and compliance issues are adhered to.

A lot of this will be realised through self-certification, which can only be achieved under the government-approved schemes for 'competent persons', which apply to companies as well as individuals. This scheme allows contractors to comply with the building regulations, micro-generation certification and a whole range of good industry practices and standards.

The feed-in tariff (FIT) and renewable heat incentive (RHI)

The feed-in tariff is a special tariff that sets out to encourage the installation of technologies for the micro-generation of electricity, such as wind turbines or solar photovoltaics. This is a guaranteed sum for every unit of renewable energy produced, which goes to the owner or operator of the technology. This could be either a commercial concern or a householder. Originally, it was set at a rate of 43.3 pence per unit (kWh) of electricity generated and was guaranteed for 25 year, but this was reduced in 2012. (At the time of writing the government is still adjusting the level of FIT, in conflict with some parts of the industry.) The FIT is paid for by electricity companies, who gain the revenue from people using conventionally supplied electricity derived from burning carbon fuel. This tariff will eventually be phased out as the balance of the market changes, because ultimately it is possible that this technology will start to replace carbon fuels universally.

The renewable heat incentive will be a similar payment scheme to the FIT for operators of systems producing heat from a low carbon source such as a heat pump or combined heat and power unit. This will also help provide a subsidy towards the payback on investment to stimulate this market as well.

Companies achieving competent person status have to be drawn from the higher end of the contracting industry. This should be a very positive step, as it will help eliminate cowboys from the industry. This one move should both drive standards higher and help to achieve the low carbon society and the elimination of bad workmanship and practice. Those contractors who view this with suspicion, frankly, have to adapt or die. This is especially applicable to those contractors who are currently exclusively fitting carbon fuel based equipment. Of course they will still have a future in the medium term, since, as a society, we are not going to get rid of these overnight, but the transition is certainly happening. We can see its beginnings in, for example, the move towards gas condensing boilers, which are more efficient. They are upwards of 94% efficient, whereas conventional boilers were about 80% efficient. However, this is merely a step, the first part of a bridge towards the low carbon society.

Contractors will eventually have to make a business decision to move completely away from carbon-based technologies. This is always dependent on the market and the legislative requirements, but its already clear that the future is electric. Governments of all political persuasions have stated that by 2016 all new housing built will have to be zero carbon. This term is confusing to some, but it means zero carbon at the point of use – and therefore no burning of fossil fuels. Will it be enforced? Time will tell, but it seems likely that all new houses will be run on electricity. This isn't necessarily zero carbon in reality, since it depends on how the electricity is generated, but it will all contribute to a significant change in the whole market, and also a change in how we view building services engineering.

Contractors, as a whole, will have to be more sustainable throughout the entire construction and engineering process. This is not just about them putting their own house in order, or about normal business processes. It's about how we conduct the entire cradle-to-grave process of building the future.

Waste

It is in the interests of contractors to minimise waste. Cutting out waste makes business sense. Yet it still doesn't happen to the extent that it should. This is because contractors pass on their waste costs to the client. However, main contractors and clients are becoming more aware of the issues, and don't want to pay for waste. This waste reduction also has the impact of reducing carbon at the same time. Reducing waste means reducing the embodied carbon in waste materials and also reducing the ever-increasing problem of landfill. The waste hierarchy (Figure 8.3) follows the same principles applied to the energy hierarchy discussed throughout this book. The two are totally interlinked through the lifecycle of resources, materials and embodied energy.

Moreover, legislation focused on maximising waste reduction will increase. This will come in the form of both carrot and stick. There will certainly be

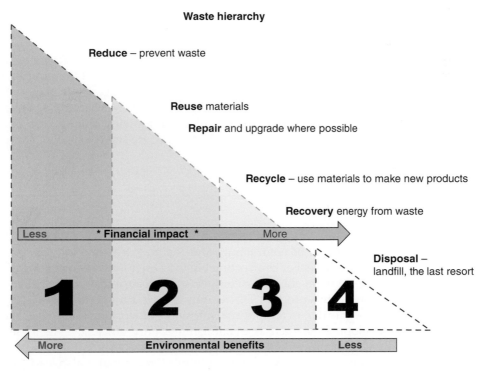

Figure 8.3 The waste hierachy

incentives for waste reduction. There is the reduction in cost of materials, plus, for any materials that can't be actively recycled, avoidance of landfill tax and disposal costs. Landfill tax will rise significantly in the coming decades. In the UK, we are quite literally running out of spaces in which to make holes in the ground. There are other escalating costs around disposal anyway – transportation, for instance. The price of oil is rising, and costs are passed on. For the present we are still an oil-based economy which has a big knock-on effect to the costs.

Water

As I stated in the previous chapter, the great Cinderella service in relation to future contracting is water. This will become an increasingly important issue. It has been neglected in the past – water saving has not been a big issue. And yet, the embodied carbon in water is significant, because we have to capture, pump and process it.

In building services, historically, water has not been managed, so what we find is that although electricity and gas are monitored and measured, water is still relatively neglected. But water use will become a pressing issue in the wake of climate change. We are starting to see this reflected in legislation. As stated in Chapter 7, for the first time in 2010, when Part G of the Building Regulations (covering water issues) was revised, it introduced a water efficiency calculator for new dwellings.

Water saving represents a great opportunity for contractors as part of diversification of their business – it is an untapped issue! Water will become one of the main issues in all of our lives as we start to see the continuing effects of climate change and growing population; and with rising water costs and increasing legislation there will be a growing market for efficiency measures.

Adding value and opportunity

The whole process of contracting, and the way it is moving forward, is all about adding value. For contracting businesses, it's a way of making sure that clients engage them above the competition, because their offer will add more value to the entire process. This is why it's vital for contractors to put their own house in order, to adopt proactive ways of working and to keep up with new developments, such as maximising energy efficiency. All these things lead to a better process and save the client money. The future is low carbon. This has the benefit of meeting mandatory targets, but it also leads to higher efficiency and reduced operating costs, as well as driving up standards. In short, everybody wins. For a contractor looking to develop more business, it provides a business opportunity for diversification. So not only are they complying with legislation, they are also ensuring their viability as a business for the future. Embracing sustainability means stealing a march on contractors that don't have this foresight.

Take air conditioning companies, for example. Under the European Performance of Buildings Directive, it's mandatory for all air conditioning systems to be inspected and certified. This is to ensure that the refrigeration gas is correct in both type and volume, but it also means that systems are certified as energy efficient. At the time of writing, it's not being actively enforced, so it may take time for compliance to become commonplace. However, this does not mean that contractors in the field should ignore the directive. On the contrary, those contractors who are ahead of the game will establish themselves in this new market with far more success than their slower competitors. These types of regulations provide opportunities for companies that are established and competent, registered and compliant. These are the organisations that will provide successful, compliant air conditioning systems – and the same holds true in other areas of construction where sustainability regulations are coming into force. The standards required can only be achieved by quality contractors who have put the time and resources into system and personnel training. This will, once again, drive up standards and eliminate the cowboys.

References

BSRIA (2009) *BSRIA Soft Landings*. BSRIA BG 4/2009 June 2009

CIBSE (2003) *Briefing No. 6*, Chartered Institution of Building Services Engineers, 2003

Egan, J (1998) *Rethinking Construction: Report of the Construction Task Force*, London: HMSO

FBnet (1999) *Continuous Commissioning – summary guide 1999*, Fit Buildings Network http://www.thefbnet.com/concom/Summary%20Guide.pdf (accessed 15.8.2012)

FBnet (2000) *The 'E-Co' Management Approach* http://www.thefbnet.com/e-co/index.htm (accessed 15.8.2012)

HVCA (2011) *HVCA Business Plus*, Heating and Ventilating Contractors' Association, January 2011

9 Main plant and building services – HVAC systems

In the past, and to a lesser extent in the present, heating, ventilation and air conditioning (HVAC) system designs have been over-specified. Over-specification of engineering is not sustainable, and leads to inefficiency being designed into the system. It had always been a convention for designers and specifiers to build in a little bit more, just in case. This has and still could be the enemy of sustainability.

Fixed thinking – assigned to the past

Part of conventional engineering practice has been that when it comes to pumps, fans and mechanical systems, many of them have been geared to fixed operating speeds that take no account of energy use or building performance. This has been due, historically, to achieving the cheapest initial capital cost. However, modern technologies can use variable speed equipment, integrated with advance control technologies, which enables the pumps and fans to operate at faster or slower speeds, automatically proportioned to the amount of energy we use. Now that longer-term requirements to comply with regulations and take account of lifecycle, energy and running costs are taking hold; a new mind-set for designing sustainable engineering will be the new best practice.

A further inefficiency occurs because HVAC pump installations often incorporate what are known as duty stand-by regimes. The idea of this is to alternate pumping systems to spread the duty and cover each other in their operation. However, it's possible to design this out by using smaller pumps running in parallel, which would effectively halve the energy use. Designers often overlook this application because they assume that in a parallel pumping scenario, if one pump fails the remaining pump will only cover half of the duty. This isn't the case because the remaining pump actually does more work – half the installed kilowatt capacity doesn't equal half the capability. In addition to saving energy and reducing the carbon footprint, two lighter and more efficient pumps are a lot easier to install and maintain. In fact, the initial cost is lower, and because you can use variable speed drives (Figure 9.1) they are a lot smaller and their electrical supply consequently less demanding.

Delivering Sustainable Buildings: an industry insider's view, First Edition. Mike Malina.
© 2013 Mike Malina. Published 2013 by Blackwell Publishing Ltd.

Figure 9.1 Variable speed drives

When looking at pumps and fans for designing, it's important to talk to the manufacturers and get their opinion and design specification because they will be able to provide much more accurate data on their systems. Manufacturers data on their components should be a starting point, rather than a fixed destination, however. An example of this would be looking at condensing boilers, which are frequently installed with system temperatures at which they'll never enter condensing mode, leading to gross inefficiency.

All of these factors need to be brought into the whole notion of design, installation and commissioning to form the basis of a proper energy efficiency strategy for reducing a building's carbon footprint. A lot more efficient designs and equipment are now being prefabricated offsite, which can simply be installed – craned in and assembled to integrate with the other systems in the building. This could be a lot more efficient in terms of time, cost and wastage. In fact, entire plant rooms are now being prefabricated offsite. All this will form the basis of the best practice for design and installation of HVAC systems, so that the system is designed for its duty without the problem of over-specification and the waste that occurs as a result of this. Some manufacturers are ahead of the game, because they want to market products that are more sustainable, so many of the pumps and fans on the market have become more efficient in their design in past decade. However, some manufacturers standardise their products and offer nearest fit, rather than matching the exact specification. An example of such a product might be the impeller of a pump. Due to using best fit, this will result in a significant waste of energy throughout the pump's lifetime, as most designers will select an oversized impeller in the pump which would, in effect, over-perform for its need. The problem is exacerbated during commissioning, because the commissioning engineer

Figure 9.2 Pumps vertically in line

would adjust the pump to reduce the pump's duty to get it down to the original design flow rate. By following this convention, the commissioning engineer in effect makes the pump work harder, and consequently more energy is used. This is impossible to rectify over the entire lifecycle of the pump itself and the associated system. The same applies to the way pipework is installed in building services. It's often better to separately model the system pipework to take account of the physical design of the building and locate pumps where possible, vertically in line (Figure 9.2) rather than placing pumps on separate inertia bases (concrete plinths). Ultimately, we want to minimise the use of resources in the entire lifecycle of the building. This would include the physical footprint and the footprint of the components, covering the pump pipework and the whole integrated system.

System design and application

There are lots of different varieties of HVAC systems. Getting the right system for the right building is dependent on a number of factors. These include the nature of the physical design of the building, both for existing and new build, and how each of these technologies can be deployed as efficiently as possible throughout the lifecycle to contribute effectively to reducing the whole life cost of the building. So much of this depends on the layout of the building, and also the physical and geographical location of the building. It will be affected by whether it has centralised plant, by the size and complexity of the systems, and by whether there is a basement, ground floor or even rooftop plant room. The key is to evaluate the best system options for the type of building. Also, is the building's physical location urban or rural? In terms of sustainability, there should always be a presumption to go for natural ventilation, but this may not be practical in a populated urban area, especially one subject to an urban heat island effect. This occurs with dense concentrations of buildings and associated infrastructure of roads and pavements, acting to store heat from building processes or more likely from solar gain. In these areas, air temperatures can often be as much as 3–4°C higher than in a rural area.

Ultimately, what are we trying to achieve with building services in a building? We are looking at the thermal comfort and well-being of the occupants of a building. So we would have to consider a variety of parameters for occupant comfort. These are divided into different groups of issues. Firstly, there are personal comfort factors such as individual metabolism, clothing and individual skin temperature. Then, there are environmental factors such as the air temperature, the surface temperature of the building structures and things such as relative humidity and the air velocity in a building. These have complex interrelations. The personal factors are, of course, dependent on the people in the building, whereas the environmental factors relate to the building type and the weather. Ultimately, the mechanical and electrical services will have to be designed and installed to take account of all of these factors.

One critical aspect which will now have more influence on the choice of specification and design will be the supply and type of primary energy or fuel to be used. Over the coming decade the move away from fossil fuels will create more specifications and opportunities for an all-electrical supply. The other options will be lower carbon choices employing combined heat and power (Figure 9.3) or heat pumps.

Figure 9.3 Combined heat and power

This will dictate the design and specification of the delivery mechanism within the building and will still be the subject of choice depending on the type of technology used. All of the HVAC systems will be influenced by these choices, but the fundamentals of the tried and tested delivery systems developed over many years will still be influenced by the perceived advantages and disadvantages, specific to the building design and layout suited to the buildings occupants' requirements.

Choice of heating systems

First we can look at the choice of heating delivery systems.

Radiators

There is a tendency in naturally ventilated buildings for radiator based systems to be specified using basic thermostatic control or individual thermostatic radiator valves, although actually the term radiator could be misleading. Radiators are actually largely convective. This depends on the design of the radiator. These systems tend to be simple and reasonably easy to install. They can provide good temperature control and create a relatively low maintenance regime, and can be designed to be fairly flexible in their layout and appearance. On the other hand, the negative side is that some designs can be slow to respond in their thermal performance. Furthermore, because of their convective nature, the thermal efficiency is not evenly distributed and can create an uneven temperature distribution in the space. They can also limit the design or layout of the office furniture, which has to avoid obstructing the heat output devices. Because these tend to be located on the perimeters of buildings, and because of the natural tendency for building occupants to open windows, you can effectively be throwing money out the window. To assist the convective effectiveness of these radiators, some incorporate fans which will be designed to improve the air movement within the space. The fans can also be of variable speed so that they can actually be used to more effectively warm the space, or heat it more rapidly. These will use more energy because of the incorporation of the fan, so although they give more efficient temperature distribution, they require more energy and are more complicated for maintenance purposes. These systems are easier to retrofit, than when using, for example, underfloor heating, which would only usually be specified in new build.

Under-floor heating

There are two systems of under-floor heating, using either a low temperature hot water pipe or an electric heating cable implanted in the floor slab or within a suspended floor. The main advantage is that they are invisible to the user and can create very flexible layout in offices and commercial buildings, especially

Figure 9.4 Destratification fans

foyers, shopping areas etc. They also allow for good, even temperature distribution throughout and don't have the same problem of stratification of temperature that you get with a radiator or convective system. If installed correctly, they can be very efficient in terms of maintenance, but have had a bad reputation in the past, mainly due to poor installation. Also on the minus side, they can be slow to respond to changes in temperature and are problematic for areas that have other underfloor services such as power and data.

Radiant heating

Overhead radiant heating is used more in industrial units, gyms, sports halls and warehouses. They are designed to provide radiant heat output in a downward direction into the heated area. The advantage of this is that it frees up floor space and, in units with low ceilings, you don't need air movement to move the heat throughout the space to be heated. The disadvantage is that it can lead to some air temperature stratification and really depends on the heat source; it can be gas fired or electric, which have positive and negatives effects in terms of the physical space. A very important design and performance issue can be addressed with the deployment of de-stratification fans (Figure 9.4) which can increase comfort control and contribute very effectively to making these systems a lot more energy efficient. The same applies to warm air units.

Warm air units

These are used in industrial applications and tend to be free-standing or mounted at a high level. These provide a lot of flexibility in warehouses or retail sheds. They can be flexible in installation for these types of building applications, although they can take up a fair bit of space if they are at floor level and, if gas-fired, they will have to be maintained more regularly. The other important maintenance and service issue is access to the units, especially those at high levels.

Figure 9.5 Modular boilers

Boilers

There are many different types of boilers, such as modular boilers, condensing boilers, dual-fuel boilers and atmospheric boilers. But all these boilers will be using fossil fuels.

Modular boilers consist of several boilers linked together (Figure 9.5). The advantage of modular boilers is giving greater control and efficiency than using one single large boiler.

Condensing boilers tend to be more efficient, because they recover waste heat from the hot flue carrying the combustion gases. They are higher in cost, but can often be worth investing in for an overall return on investment and reduction in carbon fuel use. They are an example of developing technology to burn fossil fuels more efficiently – a bridge to a lower carbon future as fossil fuels are phased out.

Dual-fuel boilers are capable of using more than one fuel, for example, gas and oil. Some even use solid fuel, which can include semi-renewable sources such as wood. These can offer greater flexibility in relation to price and market fluctuations of fuel, but need more infrastructure to supply and store the different fuels.

Atmospheric boilers are industrial boilers, which tend to be larger. These are usually boilers that use chimneys for their convection and combustion. They don't require fan assistance so tend to be quieter in operation, but are not as efficient as condensing boilers.

Ventilation

Ventilation can be provided through a number of different approaches. Ventilation systems can be split into three groups: natural ventilation, mechanical ventilation and combined systems, which use both natural and mechanical methods.

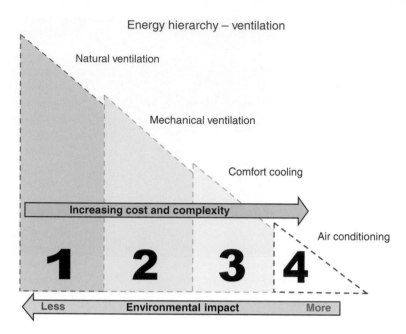

Figure 9.6 Energy hierarchy applied to ventillation

The design and specification of the ventilation system will be governed by the design of the building and its requirements. Its location will also be important – is it urban or rural? A number of issues must be taken into account, such as the physical location of the building and its orientation, particularly in relation to passive solar heat gain. The external noise levels and the level of pollution in the external environment will also be factors. The internal environment of the building and its use by its occupants will need to be considered, as will the internal airflows and the need to remove any contaminated air, replacing it by fresh air.

All through this book I have come back to the principle of the energy hierarchy. This same methodology can be applied to a whole range of building services, in examining the level of environmental impact and energy use.

Figure 9.6 shows the energy hierarchy as applied to the degrees of impact from natural ventilation to full air conditioning.

The best choice for a sustainable building, if possible, would be to go for natural ventilation, but this will be governed by all the factors just mentioned. For natural ventilation to work, it needs an easy passage for the air to enter the building through natural means, either driven by the wind or by temperature difference. It also needs a means of access or egress to create airflow, the most basic being via opening windows. This does give the user a feeling of control, but is not always practical. It can also lead to inefficient use of energy, because it's wholly dependent on individuals, and their comfort levels, in its operation. Trying to satisfy everyone's differing needs almost inevitably leads to waste unless it is combined with general awareness, education and training. This is explored further in Chapter 14.

Figure 9.7 Thermal image showing heat escape via a trickle vent

Also, it's important to consider the design of the building. If you have a large glazed area, and this isn't checked by means of shading or temperature control, this can lead to a greenhouse effect, with the building overheating due to passive solar gain. The nature of the design of the windows is also quite important – do they open fully, or do you have split windows, so that you can open the higher levels of the windows? If so, you don't end up with direct drafts or discomfort to occupants, and this creates more subtle ventilation.

In recent times, it's become standard practice – in order to comply with building regulations – to fit trickle vents These allow a small amount of air in without opening the window. The problem is that people forget to close them, and they are largely manually controlled. This means that they could be open all the time in winter, creating a draft and allowing heat loss. Figure 9.7 shows an infrared image, highlighting the problems. It's another area where people need educating about their own working environment. A technological solution has been developed, which uses a 'wick'-based, naturally controlled ventilator that responds to humidity, to open and close the aperture.

Another issue to consider with windows is: can they be micro-adjusted? Can they be altered slightly, or are they just open or shut? Also, is it possible to automate these systems? This can be done using building controls, such as sensors and actuators, with motorised opening and closing. Ultimately, fine control is a much better design feature for natural ventilation. A lot also depends on the design of the building, in terms of its security, etc., since open windows can create opportunities for burglars. Also, in a heavy rainstorm, you can't leave windows open for obvious reasons. The same challenges are created by passive solar gain – open windows can interfere with drawing the blinds. Anyway, many of the people I've observed in offices don't use their blinds to control their natural solar gain. External louvres are far more effective and are becoming far more common (see Figure 9.8). Also, in an urban area, window opening may not be an option due to noise or pollution.

Figure 9.8 External louvres mitigate unnecessary solar gain (credit: Thomas Malina)

Types of natural ventilation

A lot of people would like to opt for natural ventilation, but this is only realistic if the building structure and location allows this to function properly. If you've got good passage of air then the ventilation will flow; if not, it won't work properly.

If you can send airflow right across a building, this is known as cross flow ventilation. The air comes in one side of the building, and exits the other side. But for this to work, you have to have unrestricted airflow internally. This would work in an open plan office, but not in divided designs. In this case, you'd need a system called single-sided ventilation. This is where you have the two openings on the same wall: where you don't have a cross flow of air, you have to utilise the airflow in a semicircle, so that fresh air comes in at a lower level and takes heat/stale air out at a higher level. Figure 9.9 illustrates different types of natural ventilation designs.

The fact that hot air rises is used in stack ventilation, where the rising air goes through the building and exits upwards through a thermal stack or chimney (an example as shown in Figure 9.10) or an atrium. Another example might be a conservatory, where the openings are in the roof. It is also possible to use a combination of two or all of these methods in the same building.

Figure 9.10 shows the Queens Building at De Montfort University, Leicester. Here large stack-effect chimneys exhaust warm stale air. This effect draws in cooler air through the lower vents in the building, allowing natural cooling of the building.

So much of this depends on weather conditions because natural ventilation is, of course, a natural process, largely driven by currents created by wind.

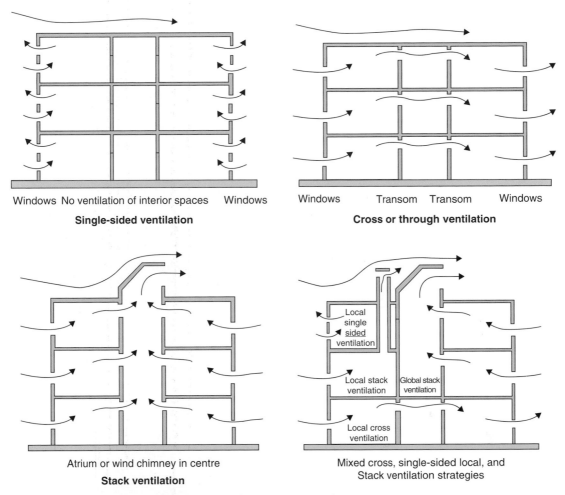

Figure 9.9 Different types and combinations of natural ventilation designs

Without that, there will be some air rise but it will be slow-moving and may be inadequate to cool the building adequately. Sometimes a combination of natural systems will enhance the functionality because if the wind is blowing a little, then you can get simultaneous stack ventilation as well as cross flow ventilation. If this is terminated on the roof of the building with a stack ventilator, this creates a very effective natural wind ventilation system. Figure 9.11 shows a good combination of high level natural cross flow ventilation and effective use of louvres to mitigate passive solar gain.

Good intentions with regard to utilising sustainable natural ventilation systems can easily be undermined by alterations to internal building structures. I have come across examples of supposedly passive systems, designed to use natural ventilation, where internal partitioning has disrupted the airflow and so later fans have had to be installed to create air movement. This was described to me as 'assisted natural ventilation', as if the energy used by the fans were somehow insignificant!

Figure 9.10 The Queens Building at De Montfort University Leicester (credit: Thomas Malina)

Figure 9.11 A good combination of high level natural cross flow ventilation and effective use of louvres to mitigate passive solar gain

In an urban area, it's less likely that natural ventilation systems will be appropriate. Not only will the area be affected by pollution and noise, but the building will also be subject to wind shadow effects from other buildings. Sustainable building designers and engineers need to explore natural options first, however, before employing mechanical ventilation, and will if necessary want to look at a combination of the two to minimise the carbon footprint of the building and create a sustainable product.

Mechanical ventilation

Mechanical ventilation systems can be divided into three groups: extract only, supply only and a combination of the two – supply and extract.

Extract only systems are employed to remove stale air from a particular process or activity. For example, they could be removing moist air from a kitchen or bathroom. In industrial or commercial buildings, they are used in underground car parks, industrial units and in fire systems for smoke extraction. Extract only systems are basic, and do not give control over the type of air that replaces the air they remove, so the new (rather than fresh) air might not be filtered or heated, for example.

Supply only ventilation systems are used to bring in fresh air to create a cleaner environment. Bringing in fresh air can sometimes be easier than using extraction. They might just be operating to create a change of air, rather than to remove air due to contamination, just like when we open a window in our living room. Fresh air has advantages, as it has the option of being heated or filtered. Supply only air can also be used in perimeter situations in buildings, where you can't open a window because of the urban area, so a fan-assisted fresh air intake provides air through a grill without the security risk.

Mechanical systems can also control the direction and movement of the air. Using mixed mode systems, which combine natural and mechanical ventilation, the natural ventilation could bring the air in – through a window, perhaps – while a mechanical system takes air out. This type of system can be used in night-time free cooling, where you extract the heat that's built up in a building by opening the window, using the ambient temperature of night-time air to cool the building down, while the mechanical extractor takes the warm air from the day out.

Larger systems usually consist of both supply and extraction, involving a central air handling unit. These comprise fans, air filtration and heating elements. The air is moved around the building using ductwork systems, which then function as the arteries and lungs of the buildings, to supply and extract the air.

The advantage of these combined systems is that they can also be used as a heating system, because the supply of air can be temperature controlled. Better-designed systems also save energy by recirculating a proportion of the warm air that's been extracted in the system. In addition, it's possible to use heat exchanging technologies, which can recover the heat from the extract air to the supply air. By passing it over a heat exchanger, the heat recovery device takes the heat from the exhaust air to heat the incoming air without mixing the two airstreams. This heat recovery is a very good sustainable building option. The main advantage of supply and extract systems is that, if they are commissioned correctly, they will create very reliable ventilation, controlling the air entering and leaving the building. If designed well, they can also be used for the night-time cooling, so they are flexible enough to be combined with a natural ventilation system.

However, these supply and extract systems have a higher capital cost of equipment and installation, and also higher running costs, as the fans for air movement will use a lot of energy to run. They also require ongoing maintenance, which includes not only mechanical maintenance but also the cleaning of the ductwork systems.

Air conditioning and comfort cooling

The challenge for any ventilation system is to control a combination of temperature, humidity and indoor air quality. A system that simply cools the air, known as air conditioning, is actually a system of comfort cooling. A true air conditioning system governs those other parameters as well – but this costs money. The more you control, the more it costs, due to the need for more mechanical services and, in the context of greater sustainability, it is vital that all the options are considered, that only systems that are really needed are installed and that the design takes into account all variables.

Air conditioning accounts for around half the total energy use of many buildings. In some buildings it is needed to provide fresh air and to maintain temperatures within a comfortable range. The shape and use of modern office buildings tends to dictate the use of air conditioning.

Typically, temperatures are maintained within a few degrees of 21°C. This corresponds with what is generally accepted to be a 'comfortable' temperature.

The more complex the system, the more effort needs to be put into design and specification, taking account of the initial costs and lifecycle costs. The operation of the system will be subject to its design in relation to the building, the capital costs and the levels of controls required, as well as the lifecycle and maintenance cost. The energy use of a full air conditioning system can account for a very significant amount of energy use within a building, and can also be a significant proportion of a building's electrical load. Full air conditioning systems will also take up a significant amount of space, both in the plant room and the runs – i.e. the ductwork. Other elements of this will be the requirement in the building for good access for maintenance purposes.

There are three main groups of air conditioning systems, which each comprise many types. The first group is complex centralised air systems, requiring one or more air handling units (AHU), which can contain heating and cooling coils. Within these systems, there are also filters and fans to move the air, and often an electric humidifier to control humidity levels in the air. The second group is partially central air/water systems, which still involve central plant, but don't necessarily need the complexity and size of associated plant such the AHUs or ductwork. Thirdly, there are local systems, which are not linked to central plant. These will be typically split air conditioning units

Figure 9.12 A wall of split condenser units

('splits') or comfort cooling. They are included because, while not strictly speaking air conditioning, they are widely referred to as such throughout the industry and most people call them air conditioning.

All these systems need to be weighed up in relation to controllability, maintenance costs, energy efficiency, their carbon footprint and their functionality and fitness for purpose in terms of air distribution. A well-designed, centralised system will have a better energy efficiency and sustainability footprint than a whole series of localised splits. I have seen a number of buildings over many years where a significant number of splits have been added or sometimes even in newly commissioned buildings, a situation occurs where the number of splits have amounted to a situation, where a lot more energy is consumed than purpose-designed central shared mechanical ventilation system. Figure 9.12 shows a wall of condenser units. This usually happens with the false economy of initial capital costs and not seeing the future picture of the comparative running costs.

Bringing it all together

Overall the key elements in the delivery of low carbon buildings will be the combination of good energy-efficient, well-planned and designed systems. These systems will need to be fully integrated to avoid system operation conflicts that have too often occurred in the past. As mentioned in the

Introduction a lot of the future delivery of appropriately specified HVAC systems will be able to be well-planned and modelled through the use of BIM and the associated working structures of a good project team. This will be further developed in the following chapters, starting with building controls (Chapter 10) which will be a vital component in making this integration of systems and practices work and in maintaining optimum energy efficiency. The commissioning process is a significant catalyst to drive this process, and this is explored in Chapter 11. From then on the continuous commissioning process will keep all the systems working at peak performance, moving in conjunction with a well-planned preventative maintenance programme to 'keep it all going' – the theme of Chapter 12 on the fundamentals of proactive maintenance.

Technology can only go so far. We then need to look at the human element, with the skills necessary to deliver these programmes and to move forward as technology develops and new technologies emerge to take on the role of delivering the lower-carbon solutions. Everyone will have to be involved as attitudes develop and change. Over time, behaviours will change, people will adapt as they have done all through history and this will happen as society and the building services industry move forward with these changes. All of these factors must be brought together in the long-term delivery of low carbon sustainable buildings, as they are to become the norm.

Further reading

A wealth of technical documentation is available covering the range of HVAC plant and services covered in this chapter. It is impossible to mention them all but a very good starting point for this chapter and a range of other issues covered in this book can be found at the following sources.

BSRIA – An extensive wealth of publications and library services. www.bsria.co.uk/bookshop/bsria-publication-list/

CIBSE – has its 'knowledge portal' which covers the complete range of building services engineering and a full range of publications looking at low carbon methodologies and practices. www.cibseknowledgeportal.co.uk

BRE – Building Research Establishment has a range of building-related publications and research. www.bre.co.uk

BRE also provide the dedicated website 'GreenbookLive', which lists approved environmental products and services linked in many cases with BREEAM and other notable scheme and accreditations. www.greenbooklive.com

10 Getting and keeping control – building energy management systems

Getting and keeping control of any process is the logical step to making sure that everything is kept to a good efficient regime. So it is vital that and building that might use an enormous amount of energy is as effectively and efficiently controlled as possible. This also goes hand in hand with providing an effective indoor climate, which is so important in creating a comfortable working environment for the building's occupants.

A building energy management system (BEMS) is an automated and computer-based system (Figure 10.1) for measuring, monitoring and managing the performance of building services that are in operation in the building, such as heating, ventilation, air conditioning, lighting and security. A BEMS coordinates the operation of various systems in the building with the aim of making sure that these are integrated and are not competing against each other. These include building services systems such as boilers, air handing units, fans and pumps. It gathers data from sensors such as light detectors or sensors that track occupancy, temperature, pressure or humidity within occupied areas of the building to create and retain a comfortable indoor environment in a cost-effective way.

Because a BEMS can control this wide range of equipment, it can make a vital contribution to increased energy efficiency and reduced operating costs for buildings. This is particularly true if information from meters for electricity, gas, water and other utilities are fed into the BEMS. In this way, the BEMS monitors how much energy is being used, and the system can reduce energy consumption by reducing the load and demand, or turning off plant and equipment that is not required. Ultimately these systems will allow easier operation of buildings because facilities management and building services engineers will have access to live data, via user interfaces including desktop PCs, laptops and wireless handheld devices. This enables them to get information on the building's performance displayed as graphs, tables of data and specific reports, as programmed and required on demand. It is also more common to monitor buildings remotely through web-based BEMS technology, for example tracking and comparing energy use and performance across multiple buildings in a company's portfolio. Specialist BEMS contractors can also provide monitoring and adjustments remotely, checking for any problems and routinely altering settings to meet changing demands made and required

Delivering Sustainable Buildings: an industry insider's view, First Edition. Mike Malina.
© 2013 Mike Malina. Published 2013 by Blackwell Publishing Ltd.

Figure 10.1 BEMS – A computer-based system (credit: Siemens)

by occupants as patterns of use change, creating differing profiles or set points for variations in the indoor climate of the buildings.

BMS becomes BEMS and can save a lot more energy

Building controls used to be referred to as simply building management systems (BMS). The history of this is quite simple – they became BEMS particularly at the time of the 1973 oil crisis, but then the 'E' for energy, fell out because energy became relatively cheap. Now, in these environmentally and cost conscious times, the 'E' is back in again with even more relevance.

Energy is in fact fundamental to the whole process. The issue is not just about controlling buildings' plant and environmental services, but very much about getting to grips with the use of energy, as a result of a building function and activity.

A good analogy, or way of looking at BEMS in everyday life, is to compare a building to the human body. To function, everything needs to be balanced and working correctly. If we feel unwell, something is out of kilter in our body, and the same is true for malfunctions in buildings. The BEMS is the brain of the building. Keeping the brain or BEMS trained and continually commissioned leads to efficiency and fewer problems for the future. Just like the body, buildings contain all of the processes to make them functional. The function os nerves and neurons is to communicate, like pipes and cables sending messages from control systems to different parts of a building.

When there is a need to change the running of plant or equipment, such as air handling units, signals are sent to devices or plant to change the running state, in just the same way as the brain sends out signals to the body if it wants to run or walk. Consequently, keeping these systems in balance is vitally important. If one of these things goes wrong, there is a big knock-on effect on the entire system. This very much relates to Chapters 11 and 12 on commissioning and maintenance. This is hugely important as BEMS require input from building operators and users as well as the automation that takes place. Building controls and building energy management systems are not 'fit-and-forget' systems: continuous monitoring, assessment and action are required to maximise the benefits for keeping good control of the building, its services and ultimately its indoor climate.

BEMS technologies are not new. The first example of an automated building control was in the 1880s. This was first devised as a bi-metallic strip, of steel and copper, which was fitted into a thermostat with a manually wound spring-powered motor. This device controlled the space temperature by adjusting the damper on a coal-fired boiler furnace. Later, in the 1890s, the first pneumatic-powered control was invented. In the 21st century, using building control systems has become standard practice. Almost all commercial and non-residential buildings have automatic building controls, coordinated by a central computer. These systems have become very complex, and have evolved very rapidly over the past two decades. Also, the software has become more user-friendly, and is able to facilitate a lot more complex control regimes to control all the systems in a modern building. Having said this, the main problem lies in the fact that the systems are so complex, and there are so many different options, that most of the advanced features of the BEMS are very much under-utilised. For example, few operators are utilising the monitoring capabilities, projections and trends of performance and energy use. These could be used to improve all of the heating, ventilation and air conditioning (HVAC) and other plant and equipment such as lighting so that energy could be controlled effectively.

About 60–80% of energy use in a building can typically be controlled by the BEMS (BCIA 2010).

It is estimated that up to 90% of heating ventilation and air conditioning systems are inadequate in some way, costing industry and commerce over £500 million per year in additional energy costs (Carbon Trust 2007).

The advantage of any new-build or complete retrofit or refurbishment of a building will be the opportunity to specify an up-to-date BEMS, very much geared to the building's needs and tied in with all of the best practice brought together via the BIM process (as covered in the Introduction). On the other hand, so many more of the projects undertaken over the coming decades will be in existing buildings. So the importance of looking at any existing systems will be the priority. The key will be how to evaluate effectively existing BEMS that may be present in a building and how to upgrade and get the best out of these systems, and then looking at ways of selecting a new BEMS if the opportunity arises. Key to both existing and new BEMS will be the importance and emphasis on commissioning these systems and making sure that the best value is obtained. Part and parcel of commissioning is the concept of continuous commissioning, which includes maintenance and keeping the continuous operation effective and efficient. The importance of developing and maintaining strategies for optimisation of BEMS is a fundamental part of this process and how to effectively integrate all the building systems for managing the buildings to gain maximum energy efficiency, utilising the BEMS.

Evaluating an existing BEMS

Many buildings will already have a BEMS in place, so the first step is to start an evaluation of the BEMS system to see if it meets the present and future needs for the operation of the building. Then a decision can be made on whether to upgrade or replace the existing system. Much of this depends on the type and age of the system, and the current state of its operability. A key decision is whether the current system can be upgraded effectively, as this will be far more economic than putting in a new system. The BEMS will not be of significant use if it is only a short-term measure. However, sometimes, driven by cost, people try to patch up a system that's already at the end of its useful life. This is a false economy. When considering a system, the question needs to asked: will it be upgradable over the following decade? It may seem like a cost saving, but it's not really economic to spend any money at all on short-term solutions. The goal is to create a system that enables the best possible energy management and control from the BEMS.

In the evaluation, there is a need to look at the energy management requirements of the building, and how the control strategies are meeting these requirements, and, in effect, does the existing BEMS meet the control requirements? In this evaluation of the current system, the questions are: are the full capabilities being used appropriately? Is the current system obsolete technologically? Can the system be upgraded effectively? Is there a good service record on the system? In fact, it's just like looking at the service record of a car. If there are gaps, then that may lead to complications with its operation because of previous neglect. Many of the perceived problems could be purely due to lack of maintenance on the BEMS. Part of the maintenance regime should be to examine the software, hardware and firmware. Has this been kept up to date? If it hasn't, then the evaluation should take into account the upgrades required. Looking through maintenance logs gives an indication of whether faults have been logged caused by plant alarms being registered because the system has been out of calibration; this could cause problems that could have been cured quite easily by keeping the system up to date and fully maintained. By asking these key questions, the facilities manager will be able to evaluate and decide on the appropriate action, whether that is to undertake an upgrade or to replace the system. The key element here in the decision-making process is to be logical and systematic, which will be key in justifying the economics of the choices that need to be made.

The actual evaluation process is vitally important. First and foremost, this needs to involve the person or persons who are monitoring and using the BEMS on a day-to-day basis, as they will be able to provide feedback and information on the positives and negatives of the system. This will also give an indication of what features need to be included in any replacement or upgrade of the system. To start an effective evaluation of an existing system, there needs to be a technical assessment which will include the design requirements, the control strategies and the physical equipment. It is also advisable to use the existing contractor maintaining the current BEMS, and it would be good practice to get a second opinion as this will add to the checks and balances on the existing contract, and ultimately act as a check on value for money and good technical operation.

Important to this whole process is the assessment of the training needs of the day-to-day operators of the system. Without training, they will find it difficult to make the best use of the BEMS. Any sort of cutting or trimming back on training on how to operate the system is a total false economy. Many of the problems attributed to poor functioning of the BEMS can be due to poorly trained operating personnel, who can waste a lot of time trying to understand a system which, if explained correctly, would be a lot simpler.

When evaluating the BEMS, this will be a good time to take the opportunity to evaluate the operation of the actual mechanical and electrical plant. Integral to this is to document fully any changes and observations that come about as part of this evaluation. The key outcome to evaluating the BEMS will be to understand the performance and cost savings that can arise from upgrading and re-commissioning the system.

Using all the factors from this evaluation, the building operator or BEMS specialist will be in a much better position to understand the system and to draw up a list of what is needed to achieve the aims and objectives of creating a better system.

Degree day analysis

An important method for checking and evaluating the effectiveness of the current controls is to use the method of degree day analysis. This is a way of using information on your building's energy use and comparing it with external temperatures. This analysis can show, for example, whether the building's heating system is operating in response to cooler outdoor temperatures, or whether it is running unnecessarily. Degree day analysis can therefore be very useful for spotting long-term energy waste, and finding ways to prevent it.

What are degree days? Degree days are used as a method of expressing the length of time and how far the outside air temperature falls below a certain threshold (known as the base temperature). In the UK, the base temperature for heating is normally taken as 15.5°C (external temperature), above which it is assumed that heating is unnecessary inside a building.

They are calculated by totalling the number of 24-hour periods over which the external temperature falls 1°C below the base temp, e.g. if it is 13.5°C for 12 hours, this is 12 hours × 2°C below base line = 24 degree hours = 1 degree day. The more degree days, the colder the climate and the greater the need for space heating.

Two factors particularly influence temperature in the UK. These are altitude above sea level (the higher above sea level, the lower the temperature) and latitude (further north it is generally colder). Local variations also occur and it is usually warmer on the west coast than the east, because of the warm sea currents. So-called 'heat islands' can also occur in large urban areas where solar energy is absorbed by the concentration of hard building and road surfaces; also, 'frost pockets' occur due to cold air sinking and collecting in some valley areas.

The following average external temperatures found over the heating season illustrate temperature differences across the UK:

Aberdeen 5.9°C
Manchester 7.4°C
Southampton 8.2°C

Variations in numbers of degree days, and therefore heating requirements, are shown by average totals of degree days over the heating season for four regions:

Cornwall 1800
London 2100
Manchester 2300
Central Scotland 2600

Using a 20-year average of degree days in the UK, numbers across the whole country range from 1650 to 3600, the highest (in the Scottish Highlands) being more than twice the lowest level (on the south-west coast).

Degree day analysis – example

These figures represent an example building that falls within the CIBSE Guide F 'Energy Efficiency in Buildings' category of prestige fully air conditioned offices. It was constructed as new during 2002/3. It has a gross internal floor area of 20,000 m², which includes office accommodation, dealer floors, auditorium, catering facilities, atria and plant rooms. The building was significantly refurbished during 2005/6. All floors are air conditioned by fan coil units. The actual gas consumption each month during one year was as shown in Table 10.1 and Figure 10.2.

With the slope of the graph in Figure 10.2 around 1500 kWh/degree day, the consumption per year can be calculated at

$$\frac{1500 \times 2051}{20,000} = 154 \, kWh/m^2/year.$$

where 2051 is the standard degree days per year used in the CIBSE benchmark figures and 20,000 m² is the gross internal floor area of the building. This puts energy consumption for heating in this building between CIBSE's ratings of typical and good practice.

However, the regression figure (R² = 0.59) shows that the relationship between degree days and energy use for heating is not strong. This indicates that heating could be better controlled to reduce its use when not required, thereby minimising energy waste.

Table 10.1 Gas consumption for each month related to degree

Month	Total (actual) gas kWh	Heating degree days (base 15.5°C)
Jan	910,245	268
Feb	704,954	258
Mar	549,496	260
Apr	574,895	124
May	499,896	109
Jun	301,789	38
Jul	244,007	31
Aug	269,478	39
Sep	273,881	67
Oct	299,861	166
Nov	424,483	248
Dec	649,589	332
	5,702,574	**1940**

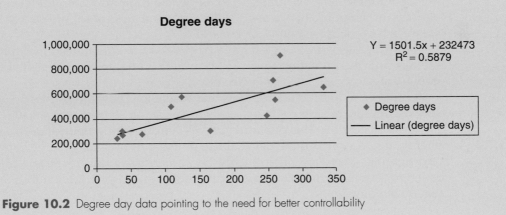

Figure 10.2 Degree day data pointing to the need for better controllability

Benchmarking – what does 'good' look like?

Degree day analysis is useful for establishing whether building services are operating efficiently, and if they are being effectively controlled. However, for any facilities manager looking to improve energy management in their building, one of the key questions must be how well their building is

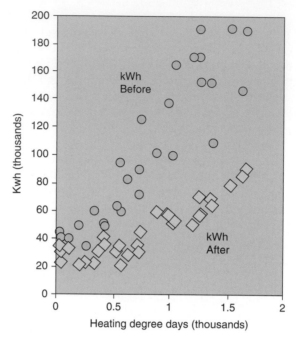

Figure 10.3 Before and after a new BEMS system

performing against others of similar construction. This is an important point for the building controls strategy too: understanding how a building could perform in terms of energy use will inform what targets should be set for areas such as energy use and CO_2 emissions.

Figure 10.3 illustrates an example of a 10,000 m² commercial building's savings from incorporating a new BEMS system with improved strategies for minimising heating related to degree days.

A number of benchmarking tools are available. The Carbon Trust offers several free documents from its website (www.carbonrust.org), including *Energy Use in Offices*. This publication gives benchmarking figures for overall energy use, as well as more detailed use on the performance of fans, pumps and lighting. There are other energy benchmarking guides available for a range of sectors, such as local authority buildings, prisons, laboratories and leisure buildings.

Defining good control strategies

One of the most important points to remember about a BEMS is that it will only function according to the system configuration. This means that if your main goal is energy efficiency, you must define your controls strategy to achieve this. It is vital to communicate these objectives to the controls professionals working with you, so that they can engineer the BEMS software according to your requirements.

There are a number of energy-saving strategies that can be considered, and discussion with your controls contractor or supplier will help to identify the most appropriate ones for you. One of the most common energy-saving controls strategies is demand control.

Simple controls can be used to turn on and off central plant and services at certain times of the day, week or even year. A more modern concept of an 'intelligent building' is to control the HVAC plant based on demands from the space rather than relying on the use of time schedules and manual control to start the plant at the beginning of the day and to turn it off once occupants leave the building. Here are some examples of how demand-driven software has been implemented to save energy.

Example control strategies

Demand controlled heating/cooling

A typical boiler plant may have been enabled via a time schedule to start at 9 am and turn off at 5 pm. Only when the outside air temperature falls below a 'hold-off' limit of say 16°C will the boilers be enabled via their time-schedules. Under these circumstances the boiler plant is permitted to operate because the outside air is cold and because the building is occupied. However, this does not necessarily mean that heating is required in the space. For example, the space may be an unoccupied meeting room. This system also relies on the engineer to set the time-schedules accurately so that they coincide with the building's occupancy. For improved control it is better to use 'demand-driven' software so that a space temperature and occupancy levels are also taken into consideration. A space temperature sensor and PIR (personal infrared) detector can be used for this purpose. This ensures that the plant will only run when the area is occupied *and* the space requires heating to meet the desired temperature level (known as the set point).

The more space temperature sensors available, the more acutely the control can be set. For example, space sensors can be deployed to take into consideration the direction of the sunlight entering a building to account for solar thermal gains. Information from the sensors can be relayed back to the BEMS so that ceiling mounted fan coil units are enabled in groups. Those nearest the south-facing windows will need to operate in cooling mode some time before those on the north side need to start at all.

Demand controlled ventilation

A BEMS is normally used to control the temperature of a building to provide heating and cooling to the space to maintain comfortable working conditions. The following example details how the BEMS can also be used to control ventilation rates within a building using CO_2 sensors.

Figure 10.4 Illustration to show damper control from demand driven BEMS strategy

Traditionally, air quality has been dealt with by setting a minimum position on the fresh air damper of an air handling unit (usually 20%). The dampers are then controlled between this 20% minimum and 100% maximum to meet temperature demands, making use of free cooling and heating where possible. The problem is that the demand for fresh air fluctuates during the day depending on occupancy levels, so very often the amount of fresh air entering the building is exceeded with this 20% fixed-open position approach. The air handling unit is therefore conditioning more air than is necessary, causing undue energy usage.

Demand controlled ventilation is used to prevent this wasteful scenario from occurring. It alters the position of the fresh air damper with reference to the number of occupants in the space. The more occupants there are, the more the dampers open (to a maximum limit), as shown in Figure 10.4.

The CO_2 sensors are more accurate than regular occupancy sensors as they monitor respiration and therefore account for both occupancy and human activity levels. The fresh air damper is modulated using the CO_2 sensor readings to achieve the desired CO_2 level. This ensures that the minimum fresh air demand is maintained at all times without bringing in fresh air unnecessarily.

Outside air has a CO_2 content of approximately 400 parts per million (ppm), but a typical office working environment will be around 600–1000 ppm. An upper limit of approximately 700 ppm is usually implemented. For accurate control of a large area, such as an open-plan office floor, it is preferable to position the sensor in the extract duct to provide an average value for the whole floor. For a multi-zoned system, however, it is more effective to have wall-mounted sensors located in each room. In this case, the sensor must be positioned so that it is not influenced directly from a single person, but must represent the air quality of the whole area.

If there is a demand for fresh air because the CO_2 levels have risen above set point, the fresh air dampers should not open fully to meet this demand, as this could have an adverse effect on the system. Consider the impact that this might have on the occupants and the heating/cooling system itself on a winter's day. The heating coils would not be able to cope with the sudden heating demand caused by an influx of cold air. Therefore, there must be a lower *and* an upper limit for minimum fresh air. The upper limit will prevent the damper from opening past the original minimum ventilation position and overloading the heating system. The demand controlled ventilation then works between these values.

It should be noted that areas with high exhaust, for example a kitchen, will require a high quantity of fresh air, and so demand controlled ventilation is not a suitable application. However, it is suited to car park extraction where carbon monoxide sensors can be used, only enabling the fans when the level approaches safety limits.

Areas where this is particularly useful are those with highly diversified occupancy patterns such as cinemas, theatres and retail outlets. It has also been shown by several studies that pupil attentiveness is strongly linked to CO_2 levels, so use of this system in education settings is now strongly recommended.

Upgrading the BEMS – the business case

A survey and assessment of your building controls and BEMS may result in the decision to upgrade existing equipment, or perhaps even to install a new system altogether. Clearly, this is a step that needs to be financially justified. A new BEMS can bring significant cost savings to a business, with reduced energy bills and more effective building operation.

A useful document for those looking to evaluate possible the possible cost benefits of a new BEMS is the British and European Standard BS EN 15232: 2012, Energy performance of buildings, impact of building automation, controls and building management (BSI 2012). This document is based on research carried out to support the European Energy Efficiency of Buildings Directive (EPBD). It demonstrates the link between different generic types of building control and the level of energy saving that can be achieved in different buildings such as offices, schools and retail outlets.

For specifiers, BS EN 15232 assigns classes A, B, C or D to levels of control within a building (Figure 10.5), and shows the resulting energy efficiencies that could be expected. This is an invaluable tool for those looking to balance capital investment against long-term energy savings. It also makes specification of BEMS a much clearer process for everyone involved, from the end-user client to the installer. Table 10.2 is taken from this standard, and shows the potential impact of controls on the thermal efficiency of non-residential buildings.

The BCIA recommends use of BS EN 15232 for specifiers and end users who are planning a building controls or BEMS project. Not only does the standard give a clear indication of energy savings that can be expected from the use of controls, but it also offers a common language for specification.

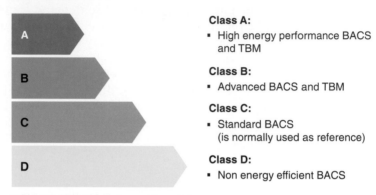

Acknowledgement to

A guideline for control – EN 15232 Building Automation – impact on energy efficiency

BACS Energy performance classes

Class A:
- High energy performance BACS and TBM

Class B:
- Advanced BACS and TBM

Class C:
- Standard BACS (is normally used as reference)

Class D:
- Non energy efficient BACS

BACS – Building Automation and Controls System
TBM – Technical Building Management Systems

Figure 10.5 A guideline for control – EN 15232 Building Automation – impact on energy efficiency

Table 10.2 The potential impact of controls on the thermal efficiency of non-residential buildings

Building type	Building control efficiency factors – thermal			
	D	**C**	**B**	**A**
	Non-energy efficient	**Standard (reference)**	**Advanced energy efficiency**	**High energy efficiency**
Offices	1.51	1	0.8	0.7
Schools	1.2	1	0.88	0.8
Hospitals	1.31	1	0.91	0.86
Restaurants	1.23	1	0.85	0.68
Retail	1.56	1	0.73	0.6

Specifying a new BEMS

Installing a new BEMS presents an opportunity to implement a controls strategy from the start. Elements to consider are as follows.

Objective of the controls system: what is your main strategy for the BEMS? It could be energy-saving, occupant comfort, maximising use of on-site renewable systems, or a mixture of all these.

Figure 10.6 Specialist BEMS contractor

Selection of the contractor: It is important to appoint a controls partner who can work with you to implement your strategy, and also in the long term to help with maintenance and continuous commissioning (Figure 10.6).

Specification method: The three main types of controls specification are standard guide specification, performance specification and proprietary specification. Standard specifications are widely available and generic. Performance specifications are determined by the particular requirements of the client. Proprietary specifications are more exacting as they allow for the selection of particular building controls products. The choice of specification method will depend on requirements and budgets.

The main objective is to get a system specified that works first time, as any alterations during or after the project will add to the cost. Any new system should be specified with maximum opportunity for expandability and easy upgrading. Look to include built-in upgrades for all software that will not involve the purchase of more hardware. This should include the ability to upgrade without changing the main user control interface, and leaving available enough capacity on the system to cover any need for renovation, upgrade or additions to the system.

Interoperability – open systems: Specifying systems or components that are compatible with other systems is always a good idea. Systems such as BACnet, KNX or LONworks are examples of protocols that many BEMS manufacturers use. The competent vendor or contractor will be able to advise on the appropriate use of these systems.

Continuous optimisation

As mentioned at the start of this chapter, building controls should not be treated as 'fit-and-forget' elements of a building. Continuous optimisation is a process that should be central to your controls strategy because it means that the building is maximising energy efficiency in the long term, and you are reducing the likelihood of faults going undetected through continuous performance assessment.

Monitoring is the first step. At its simplest, this means simple status monitoring: Is equipment running when it is supposed to? Is it off when it should be? Does it open/closed appropriately?

The next step is to gather information on the energy use of the building. From this it is possible to identify areas of energy waste, or to spot areas around a building where energy use is unexpectedly high. The reasons for this can be very simple. For example, you may find that a sensor requires repair, or even that someone has positioned office equipment such as a photocopier under a sensor. A quick visual check may be all that's required to solve these problems.

Deeper analysis of building energy use can include comparison of your energy against industry benchmarks. At this point it can be useful to implement energy-saving strategies such as demand-driven ventilation or heating. You may also find that simply encouraging building occupants to think more about their energy use can lead to excellent results.

The results of such strategies and campaigns can then be further monitored for their impact on energy use, using degree day analysis, and the process

Benefits of continuous optimisation

Reduce energy and operating costs
Constant comfort level for occupants
Increased reliability and efficiency of building services equipment
Extended life of building services equipment
Lower the impact of your business on the environment
Produce 'provable' energy efficiency figures
Help to increase the competency and understanding of your operational personnel

begins again. The true value of facilities managers is that they can track the success of energy-saving schemes, highlighting where they have saved money for the business.

The BEMS needs user involvement

One of the most important points to bear in mind when specifying a new BEMS, considering an upgrade or even running your system day to day is that the system will only perform to its maximum potential with user involvement.

When specifying a BEMS be clear about user requirements in the short and long term. For example, if energy efficiency is the priority, then this should be stated clearly in any specification documents and in discussions with contractors.

One of the main benefits of a BEMS is that it can supply a wide range of in-depth data about the building's performance. While it may be tempting to ask for as much data as possible, you should consider carefully how this data will be used, and how it will contribute to efficient building operation. For example, it is possible for a BEMS to track temperature changes to air or water supplies every 30 seconds if programmed to do so – but is this strictly necessary? Gathering large amounts of unused data can create unnecessary analysis for the facilities team which will not necessarily yield better results than if data is collected every 30 minutes, for example.

It is also important to ensure that all the relevant staff understand how to use the BEMS. As staff leave and others join the organisation, it is important to ensure that the facilities team understands how to get the best out of the BEMS. An ongoing training programme is the best solution, along with good documentation. Although the BEMS is a highly sophisticated automated application, it is as good as the people who operate and manage it. BEMS is a mix of different technologies, but above all it needs people to operate and understand it. This combination is essential to the delivery of a low carbon sustainable building.

References

This chapter is based on the British Institute of Facilities Management *BIFM Good Practice Guide* jointly sponsored by the Building Controls Industry Association (BCIA) and the Federation of Environmental Trades Association (FETA). Written by Karen Fletcher and Mike Malina due for publication during 2012.

The BCIA produces numerous articles on the use of controls, and these can be found on the BCIA website at: www.bcia.co.uk

BCIA (2010) Building Controls Industry Association http://www.bcia.co.uk/files/ John%20O'Leary.pdf (accessed 20.8.2012)

BSI (2012) BS EN 15232:2012 Energy performance of buildings. Impact of building automation, controls and building management. British Standards Institution, February 2012

Carbon Trust (2007), *Building Controls* – CTV032

CIBSE (2004) *CIBSE Guide F, Energy efficiency in buildings*; available via: www.cibseknowledgeportal.co.uk

11 Commissioning and handover for energy efficiency

The whole activity of commissioning is much misunderstood within construction. Too many people in the industry think of it as something that happens at the end of the construction process. This, in essence, is why things can go drastically wrong! In the course of consulting work, I come across many buildings that are not performing as intended simply because they were not commissioned properly.

What tends to happen is that the construction processes overrun due to other contracting pressures or due to delays and mishaps during the process. Then, although perhaps originally the commissioning period was intended to run for several months, because the main contractor and/or the client wants the project finished, there will be an attempt to get the commissioning done in a third of the time originally allotted. This leads to rushed or half-hearted attempts to get things working first time. Any process that's rushed always leads to inaccuracies and, with mechanical and electrical issues, it leads to services that just aren't performing properly.

Despite this, over the first decade of the 21st century, the profile of commissioning has risen significantly, and it now even has its own trade body, the Commissioning Specialists Association (CSA). I would definitely recommend using a company that has a specialism in commissioning and is a member of the CSA.

The Commissioning Specialists Association (CSA) was formed in 1990 by a group of the Britain's leading commissioning companies.

It is an association for the commissioning industry within the construction and building services engineering industry. The membership comprises commissioning industry companies, individual commissioning engineers and associated companies, including equipment manufacturers and instrument suppliers who have a vested interest in ensuring that the commissioning role in today's complex buildings is undertaken to a uniformly high standard.

The CSA's main objective is to offer anyone who utilises the services of commissioning companies and engineers the guarantee of a professional service, based on trained, qualified and experienced field personnel, backed up by a quality of service underpinned by adherence to the CSA's aims, objectives and code of practice.

Delivering Sustainable Buildings: an industry insider's view, First Edition. Mike Malina.
© 2013 Mike Malina. Published 2013 by Blackwell Publishing Ltd.

In addition, the CSA acts to promote the commissioning industry, putting forward the views of its members across the building services sector and wider construction industry. A major role is keeping the CSA membership informed of developments in technology, equipment and instrumentation (Figure 11.1) and changing legislation which has implications for the way the commissioning industry functions and operates.

The CSA has placed a lot of resources and effort into training and career development for the commissioning industry, developing a fully structured package to encourage commissioning engineers to follow a defined course of study to develop their theoretical knowledge and practical skills, thus enabling them to fully realise their potential in their chosen career. The aim is to ensure that all commissioning staff have the necessary skills to be able to carry out their work to the highest professional standards and provide assurance of a reputable image for both company and individual benefitting in repeat business (Figures 11.2 and 11.3).

The main benefits are that staff provide a consistently high standard of work and commissioning companies are guaranteed to provide clients with an excellent commissioning service.

www.csa.org.uk

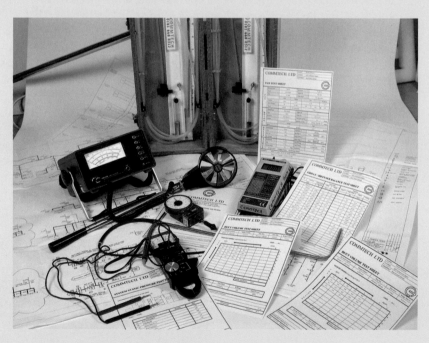

Figure 11.1 Specialist commissioning instruments, equipment and documentation

Figure 11.2 Commissioning training – example, water balancing and flow rate measurement

Figure 11.3 The importance of hydronic balancing can be practised on a purpose-built system

What is commissioning?

Commissioning is the activity that ensures that the building performs according to the original design intent, to suit the needs of its owners, operators and occupants. Buildings are very complex, and becoming more so with modern building technologies, so the process of commissioning has evolved to cover integrated building services, especially where mechanical and electrical services are concerned. Where there are so many different systems, this complexity brings its own dangers. Unless a building is being built to an exact model and specification that has been replicated elsewhere, we can in effect say that every building is a prototype. For instance, heating and cooling systems can be quite literally competing against each other if they are not integrated and commissioned properly.

Commissioning should not be a one-off activity that occurs only when a new building is completed or an existing one refurbished. The way buildings are used can have a dramatic effect on the performance of the various systems operating in a building.

Now that more complex shared systems have evolved, these problems can be exacerbated by the natural churn that takes place in the average office environment. People move office partitioning around, for instance, which can lead to situations where entire offices are without either air conditioning or ventilation – or both! I have, in the course of survey audits and inspections, found air outlet vents outside the room where they were originally intended for. Successive office reorganisations can lead to chaos in the delivery of building services, as sometimes the building services seem to be missed!

Problems also arise through people's natural tendency to adjust heating controls. Thermostatic radiator valves are a prime example. They are generally marked as 1–5, and each setting corresponds with a set temperature. They are set to 'switch' off and on as the temperature rises and falls. People don't always understand that, and use these as an 'on-off' which overrides the automated function. Many people do not realise that if they set the radiator to 5, the hottest setting, they will tend to overheat. When that happens, however, they then open the windows. If the setting is set to 'off', on the other hand, and someone has put furniture in the way so people can't get to the valve, people get cold and may bring in a supplementary electric heater. I see this on a regular basis, when conducting building energy audits. These are just some of the reasons why commissioning is not always as successful as it should be. This is mainly due to the lack of awareness of the occupants about how to actually control the system settings in their own working environment. This is the type of action that could be addressed in a process such as soft landings discussed in Chapter 8.

Commissioning, in short, becomes more complex over time because after people have adjusted the controls; this can have a major knock-on effect on all the other systems. If something leads to overheating, then perhaps the cooling can also activate to counteract this. People then fiddle with the controls again

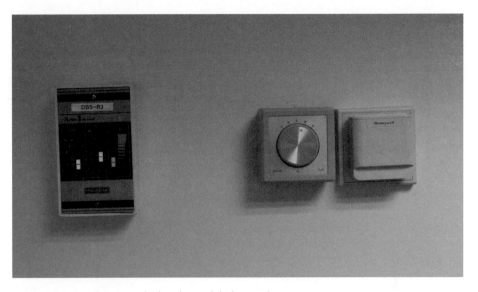

Figure 11.4 What controls do what? – labeling and awareness is important

(Figure 11.4). Sometimes, they can end up with the heating and cooling full on because there's no indication of exactly what the controls actually do. Worse still is when they open the windows as well. This is how systems end up competing against each other – they can end up using three times as much energy as necessary. Not only is it bad for the indoor climate and environment, it's also tantamount to throwing money out of the window. This leads on to the real need to make sure that systems are integrated and controlled as discussed in Chapter 10. It is also vital that people are involved and are sufficiently motivated to take charge and act to make sure that their own working environment is operating correctly.

The commissioning process

The commissioning process needs to be a staged process following a logical programme order. It is a process that lends itself to fitting in with the growing trend for using BIM (as covered in the Introduction) and all the occupancy issues post building handover, such as the soft landings process as referred to in Chapter 8.

Also referred to in Chapter 8 was the pioneering PII project which developed an early version of the continuous commissioning process defined at the time as the E-Co management approach along with the concept of a designated E-Co role for one person on the project management team (FBnet 2000). I tested this process on three successful projects between 1999 and 2002.

This method was further proposed to be integrated with the RIBA work and design process as below.

Lifecycle Development Stages (FBnet 2000)
(after RIBA Design Stages – *additions in italic*)
a – Source of materials
b – Processing & Manufacture
A – Inception
c – Principles of development, including design, construction, operation and end of life (to be used for continual reference throughout project)
B – Feasibility
C – Outline proposals
E – Scheme design
F – Detailed design
F/G – Production information
H – Tender action
J – Project planning
K – Operations on site
1 – Commissioning
L – Completion
2 – Operation and Maintenance
M – Feedback
3 – Fine tuning and minor adaptation
4 – Ongoing Recommissioning   *[Does building meet business needs? Yes, go to 2]*
                                                  [No, go to 4]
4 – Refurbishment (& Re-commissioning)   *[go to a]*
5 – Feedback II
6 – End of Useful Life
7 – Reuse (& Re-commissioning)   *[go to 1]*
8 – Decommissioning
9 – De-construction
10 – Recycling or disposal
11 – Feedback III

When I was working at Commtech, a detailed commissioning management process involving a set of sequenced stages was developed (Figure 11.5), creating a process for a commissioning audit trail leading to the project's completion and laying a sound foundation for the continuous commissioning process for the future of the building.

The actual process starts with the initial design and should then be followed by a design review, looking at specific issues of systems integration and a focus on how this will relate to the commissioning process.

Next, there is the physical installation of the main mechanical and electrical services, including the BEMS controls. This is followed by the process of pre-commissioning, testing, balancing and inspection. Examples of commissioning testing and validation are shown as Figures 11.6, 11.7, 11.8 and 11.9.

It is important at this point to make sure that third-party requirements for other systems and process is sorted as part of the commissioning management process. This is followed by the actual commissioning, which includes

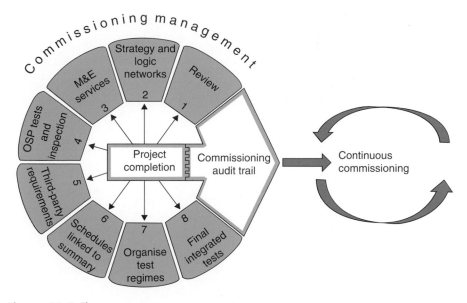

Figure 11.5 The commissioning management process

Figure 11.6 Commissioning – on-site validation and measurement of flow rates

Figure 11.7 Commissioning inspection and validation – check of pressure gauge and airflow

Figure 11.8 Testing, measuring and setting airflow

providing readable and understandable documentation for feeding into the operation and maintenance manuals and ultimately the building log book. Finally there is the ongoing process of continuous commissioning and the maintenance programme, that will keep the building systems in good operational and efficient state.

Figure 11.9 Testing and validating water flow on large chillers

The individual systems throughout this process should be commissioned to give maximum energy efficiency. A key part of this process is to make sure that commissioning specialists are used. They can provide the proper testing and commissioning that is required to commission all the building engineering facilities. It's vital to put a lot emphasis on getting all the systems working together. Good commissioning takes a holistic view, and works for integrated and interdependent systems. This is the chance to get the building set up correctly first time, with appropriately skilled engineers and technicians. There is unlikely to be an opportunity again in the life of the building, where the correct people are able to put the building through its paces without it affecting the people in occupation, unless the building is empty at some point in the future.

A more detailed programme of commissioning management is outlined as follows, using the stages of design, the design review and installation through to handover and the putting in place of a continuous commissioning and maintenance programme.

Design

To make the building work, we have to consider commissioning right from the project's conception, even before the design stage. This is crucial. It is the beginning of the process which leads into the design and on into the construction process itself. To make this work, the client, property developer, building owner or facilities manager should appoint a commissioning engineer, right from the design concept stage. Unfortunately, the importance of this isn't understood and very often real opportunities are missed. Ultimately, the buildings that have commissioning built in from the start and are commissioned correctly will operate more effectively and be a lot more energy efficient. Too often, a commissioning engineer or manager is seen as a luxury or, at best, an additional expense. Despite this, the savings from a proper commissioning process will far exceed its initial costs. In the long term, over the life of a building, additional running costs will amount to far more than the initial expense of employing a specialist manager to start with. Surely, from a business point of view, minimising the inefficiencies, snagging and identification of ongoing defects during the construction process makes economic sense for everyone. I'm hopeful that the BIM process will add weight and importance to make this happen a lot more in the future. Ensuring that there is a professional commissioning process with a professional manager is also more likely to get the building delivered on time. At the design stage, all involved should not just be focusing on the design – they should be looking at the entire process. In construction, you have to think laterally about what's going to be happening in a month's time. Commissioning is one of those things that needs planning, and it's important to be constantly looking ahead. You have to build everything into the design phase.

Design review

Normally, within the design review, the design process is the responsibility of the architect or designer, who is more concerned with designing and specifying the building structure – and its finishes – than they are with the building services. They are liable to only consider building services later. However, the mechanical systems will be dependent on the materials and structure of the building. The commissioning manager or engineer is normally responsible for making everything work in the construction process itself. Their role is about making sure services are commissioned and tested properly. It's often the case that they're not involved at this early stage, but it's important that they are. If the designer was able to consult a commissioning manager regarding the appropriateness of their design at an early stage, this could add a lot of value. Problems could be nipped in the bud. That is the key to a successful future project. The important part of this design review is to make sure that the commissioning process as a whole is controlled and placed under the supervision of a single person or team. The key is that the commissioning manager should be independent, reporting to the client, but also working with the project team (Clark 2003). The key element of the design is to keep the system as simple as possible; too often, it's over-complex. The best buildings are designed for continuous commissioning without difficulty. We need to design out complex commissioning on site by using more pre-commissioned services – for example calibrated and engineered off-site mechanical and electrical services. At the same time as the design is created, all the stages of commissioning need to be put into place as part of the design. This should include easy access to all systems for calibration and testing, which means being able to get at them without dismantling other parts of the system We should design out inefficiency and unnecessary obstacles for a future efficient maintenance programme.

The type of commissioning also has to take account of future maintenance. Part of organising the arrangement of plant and equipment in the plant room is ensuring that it can be maintained for the future. There should be adequate provision of testing ports and valves, and flushing facilities for wet systems. This should enable easy maintenance. On the other hand, if this is not designed in at the beginning, this will be creating trouble for the future. Regarding flushing, there is a need to make sure that there is adequate facility or provision to enable water treatment to be adequately carried out.

Part of the design should also take into account interoperability of all the systems, particularly with regard to looking at how the control strategy is defined. Again, part of the design process will be looking at the complete situation of operational efficiency. If you are looking at a design, and doing a review, you have got to ask whether there's a better way of doing things, taking account of the fundamentals regarding the building's function. If you know what the function and operational intent are going to be, then you can design the systems adequately, including the commissioning element.

As part of the specification for looking at plant and services, there needs to be provision made in the design to fully understand and predict the full working capacity of all the systems. There is a need to make sure that the plant running at maximum load is fully tested, even if it's never going to get to that stage. It's a good opportunity to put into place start-up and operating procedures, and to pre-document them before the creation of the full documentation and the development of a continuous commissioning strategy. If the commissioning manger or engineer can do this as part of the design review early on, it will lead to a much more efficient building process. Significant time and money will be saved, and the procedure specified for the project information flow to make sure things are done right first time, minimising costs for rectifications or changes during the construction project. (It's sometimes worth doing part load testing, as the building is more likely run at part load for the majority of its life, although sometimes controls have difficulty in coping with this exercise.)

Installation

A key part of developing the programme for installation will be time dependent, and will hinge on the wider project programming and planning. Every activity has to be timed, phased and mapped out. The commissioning programme needs to be integrated within the wider project plan, which shows the sequence and interrelations of trades and plant that form part of the construction programme. Throughout the process, all members of the construction programme should be made aware of the need to allow for full and ongoing commissioning and testing.

Commissioning is an ongoing process, and each stage of installation and commissioning has to be fully witnessed and documented. There must be no mistake about what has taken place. Every element has to be proven to be

Figure 11.10 Blower door – room air pressure testing

installed correctly and to be working to optimum efficiency. When drawing to the close of the installation phase, a fundamental part of the commissioning programme is to do an integrated systems test (IST). This is to make sure that all the systems that have been tested and witnessed individually will actually work together. Further to this, there has to be an indoor climate environmental test to check all the environmental parameters – temperature, humidity, noise level, lighting adequacy and specialist testing, for example building air pressure testing to show the building fabric integrity for airtightness (Figures 11.10 and 11.11), to make sure that all fulfil the building design intent and that they fall within design operational parameters. Figure 11.12 shows an example of highly specialised commissioning for clean room validation, mainly focused in the pharmaceutical and electronic industries.

Every construction project is different, but whether it's a refurbishment or new build, the installation sequence is the same. There is a logical sequence of events that can be checked and signed off. People witness the installation and attest to fact that it works.

Handover – continuous commissioning and setting up maintenance

It's at this point that the commissioning process sets up everything for the future with regard to the maintenance regime (Chapter 12). This point of the commissioning process is crucial, because what can happen is that when the maintenance regime starts to happen over the lifecycle of the building, you find that the maintenance company starts to report that plant and equipment weren't commissioned properly. This is something to avoid at all costs, so commissioning and maintenance need to be fully integrated. After all, continuous commissioning and planned preventative maintenance can

Figure 11.11 Testing for whole building airtightness

Figure 11.12 Specialist commissioning – cleanroom validation

run well together. To organise and record these processes, we need the building log book, which will be the single document with all references to operation and maintenance manuals (O&Ms), the original commissioning test sheets, the design and operation strategy for the building and the continuous checks that need to be made. For commissioning to succeed, the right people need to be involved, which is why so much emphasis is placed on delivering specialist training by the Commissioning Specialists Association (Figure 11.1).

Regulations and standards for commissioning

At long last, the commissioning process and sign-off has been detailed in the latest revision of the building regulations and mentioned in the approved guidance document – Part L.

The Commissioning of Fixed Building Services is covered in Regulation 40 and 44 and referred to in Approved Documents L2A and L2B Part L 2010, specifically stating: 'Reasonable provision for commissioning is to prepare a commissioning plan based on the guidance in CIBSE Commissioning Code M. Commissioning should be checked against that plan. Commissioning and testing of building systems is to be signed-off by a suitably qualified person and a commissioning certificate supplied to the building control officer'.

Ductwork commissioning should be undertaken as described in HVCA DW/143 document (HVCA 2000).

References

Clark (2003) *Achieving the Desired Indoor Climate, Commissioning*. The Commtech Group, Studentlitteratur, Lund, Sweden

FBnet (2000) The 'E-Co' Management Approach http://www.thefbnet.com/e-co/index.htm (accessed 21.8.2012)

HVCA (2000) *A practical guide to ductwork leakage testing* ISBN 978 0 90378 3309

CIBSE Commissioning guides
   Commissioning Code A: Air distribution systems (1996/2004)
   Commissioning Code B: Boilers
   Commissioning Code C: Automatic controls
   Commissioning Code L: Lighting
   Commissioning Code M: Commissioning Management
   Commissioning Code R: Refrigerating Systems
   Commissioning Code W: Water distribution systems 2010
   (available from the CIBSE knowledge portal: www.cibseknowledgeportal.co.uk

CIBSE and BSRIA work together on a number of commissioning guides and codes. These are also available on the CIBSE Knowledge Portal and at BSRIA: http://www.bsria.co.uk/bookshop

The Commissioning Specialists Association (CSA) also publishes a range of specific industry training, standards and guides: http://www.csa.org.uk/publication.html

12 Keeping it all going – the importance of maintenance to sustainability

We sometimes think of maintenance as a manmade activity, but if we look at the natural world around us we can see that maintenance is an integral part of nature's systems. All of nature's processes are based around maintaining all of its systems. The body repairs itself; if we cut ourselves, the healing process begins almost immediately. Doctors assist natural processes, but had we not evolved to repair ourselves, their work would be impossible. In fact, doctors and engineers actually use similar diagnostic tools and equipment. Thermography, ultrasound, stethoscopes, even endoscopes are used by both professions, for instance. In the modern world, we have come to understand the importance of maintaining ourselves and our health. And yet, it seems, we have some way to go when it comes to understanding the importance of maintaining both the natural and built environment.

Maintenance is fundamental

Maintenance should not just be an add-on – it's fundamental. When considering the sustainable built environment, more emphasis is placed on the design and construction phase than on the day-to-day running of the building. This is a factor that can detract from long-term sustainability. Our aim should be to minimise the whole lifecycle impact on the environment and resources that are needed.

According to the Royal Institute of Chartered Surveyors (RICS 1990), 'The design and maintenance process in the construction industry needed to be more closely allied as in the motor industry where design and subsequent maintenance frequently have an equal consideration.' They also maintain that early consideration of maintenance and early discussion about this can pay dividends in terms of the ongoing costs of the building through its lifecycle. This is all very well and good, but it's often difficult to realise these ambitions against the backdrop of prevailing commercial realities. People are driven by the price tag of upfront costs, and instinctively work to minimise these. All too often, the fact that this will create higher costs further down the line is something that is not considered. So the links with both the concepts of BIM

Delivering Sustainable Buildings: an industry insider's view, First Edition. Mike Malina.
© 2013 Mike Malina. Published 2013 by Blackwell Publishing Ltd.

and the BSRIA model of soft landings can play an important role in taking maintenance issues forward (see the Introduction).

Often maintenance and the concept of continuous commissioning have taken a back seat, but maintenance and keeping the building operating is vital to the long-term sustainability of the building. The key maintenance regime should be planned preventative maintenance (PPM).

PPM is a planned system operated over a given time period, where maintenance activities are scheduled in an organised fashion to systematically check the function and operation of building services plant and equipment. It sounds routine, but, like many routine activities, it's vitally important. It includes things like cleaning and lubricating, replacing component parts and changing filters. After all, when you think about all the equipment in building services, there are an awful lot of moving parts: bearings, motors, fans – some spinning for thousands of hours – as well as air and water being drawn and moved around and being heated and cooled, so they all have to be regularly checked.

This kind of programme will reduce money spent on repairs and unscheduled shutdowns; so adding to your maintenance regime will extend the life of the plant and equipment, and therefore the sustainability of the building. It also adds to the management of the production of embodied energy. Less maintenance will lead to more energy use, impacting on the lifecycle costs of the entire process. Therefore, the operational maintenance regime is an integral part of the sustainability and long-term efficiency of any building.

Basically, it's about the difference between proactive and reactive maintenance. Conventionally, most maintenance happens when something breaks down, and it is therefore reactive. There's a tendency to adopt a 'wait and see' approach – which is effectively saying, 'Let's see what happens and react when something falls over'! This is a bad system, because it has many hidden consequences, and it certainly doesn't tie in with sustainability. There is an inextricable link here between many important factors. It's important to realise that PPM, sustainable building services, energy management and finance all need to be considered together. We need a holistic approach.

I believe it's important that the plant room isn't hidden away. It's easy for people not to appreciate the importance of its function to the building. One strategy where possible is to make it visible as shown in Figure 12.1. This would raise the profile of maintenance as well as building services engineering in general.

With reactive maintenance systems, there is a much greater risk of complaints from building users as there is more likelihood of building services breaking down. Also, when things don't work and there is more down time, there is usually also more consequential lost productivity. This is very important to remember: if maintenance is poor or not properly planned, there is a silent creeping loss of performance. It's like a car – if you don't keep it tuned and correctly maintained, you will end up stuck by the roadside. Sometimes, everything can still keep going, but if you don't put air in the tyres on regular basis, for instance, then there will be an inevitable loss of performance.

As with the car, so with buildings. It's important that checks are not only performed on plant and equipment. In a building, for example, heat exchangers and the air handling and conditioning systems, and refrigeration plant all

Figure 12.1 A ground-floor plant room with a viewing panel for all to see

Figure 12.2 Blocked filter – bad for operation and energy performance

need regular maintenance. Fins should be vacuumed and cleaned, otherwise surface dirt will inhibit functionality. There need to be checks for corrosion also. With packaged air conditioning, filters should be cleaned or replaced on a regular basis as part of a planned maintenance schedule. Figures 12.2 and 12.3 illustrate these points.

As discussed in Chapter 10 the BEMS is a key component in controlling the performance of the building as well as providing energy savings. The BEMS also has a considerable bearing on the PPM regime. For example, even though a documented PPM regime may specify a change of filters regularly, say every 3 months.

Figure 12.3 External refrigeration condenser unit fins need cleaning

Case Study – maintenance pays

A ventilation system which operated to remove stale air from a leisure centre swimming pool operated on a 24 hour, 7 day a week basis and featured two-speed fan motors which would change speed depending on the internal environmental climate of the building. Because of the lack of maintenance the passage of air was badly blocked by dirty filters and the fan motors were consequently operating at full speed all of the time. Once the heat exchanger was cleaned, the free passage of air allowed the system to operate on low speed for most, if not all of the time.

The large fan motors on this system were rated at 15 kW. After maintenance cleaning, a lower speed usage was enabled with a less restricted air path. This resulted in a saving of approximately 5 to 6 units of electricity per hour (6×24 hr $\times 365$ days $\times 8$p kWh (unit) = £4205). The cost of maintenance was £230, which resulted in an incredibly quick payback.

This provides an example of a not-too-uncommon case that gives overwhelming support for the absolute need for planned maintenance for building services equipment. The financial savings speak for themselves, as well as the direct reduction in energy use and subsequent environmental impact. The equipment wear is reduced, thus maximising life expectancy and reducing the potential of unexpected breakdowns, helping early fault recognition and giving big improvements in the hygiene control and a better quality indoor climate for the building. The same applies to water treatment – scale in pipes could lead to significant loss of performance.

The BEMS could tell us how many hours an AHU has actually run for. Also a sensor linked to a differential pressure (DP) switch can measure the airflow performance across the filters and will tell you if the filter is actually dirty with a consequent loss of performance and is in need of replacement. It is unfortunate that sometimes even though a PPM regime is put in place, it becomes prescriptive and I've known of some buildings where a contractor has to change plant or fan coil unit filters every 3 months on vacant floors because a PPM sheet said so! The same situation has also occurred on the maintenance and longevity of lighting consumables. A group of fluorescent light tubes may be scheduled for replacement every six months, yet many may not have been in operation in specific areas for different reasons. Yet this may be ignored and they are changed anyway. It is again possible to log the actual usage via the BEMS or on the energy measurement, so that a more efficient and accurate PPM regime could be carried out. So it is important that common sense is applied and a re-evaluation of any PPM regime is essential to avoid this sort of situation occurring.

Is it worth the risk?

You can quantify the relationship between the risk of plant failure through lack of maintenance and the consequences and impact on energy performance in a simple graph (Figure 12.4). This can relate the likely failure rate of the building services plant on one axis to the consequences of action or inaction on the other axis. This matrix could apply to many things, such as health and safety, for example. Essentially, we are showing the consequences of failure to invest in maintenance in relation to energy performance and maintaining a sustainable building. Put another way – and perhaps this is why the importance of maintenance is poorly understood – it is often related to the slow and silent, creeping loss of performance that may often occur, as opposed to the obvious signs of when something actually goes wrong or clearly stops working.

Everything discussed in this book is ultimately interlinked. The work of the contractor, controls and commissioning specialists, all tie together with the efficiency and energy use of the building. So, if we know all this, why do facilities managers and organisations cut their maintenance budgets? The short answer is: because it's easy to do. A budget of £100k could be reduced to £75k at the

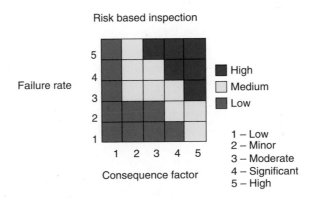

Figure 12.4 Maintenance risk matrix

stroke of a pen. In cash-strapped organisations, it's all too tempting to do, but it's far too simplistic to be a robust solution to any issue. The problem is that the decision makers often don't have the knowledge to understand the consequences of their chosen course of action. Beyond what the facilities manager can influence, senior people often make decisions with really damaging effects. Elements of risk and luck are involved in these choices. Through the luck of the draw, sometimes no obvious or immediate impact is seen. However, over the longer term, there will inevitably be problems.

This is why it's absolutely crucial that energy managers and building services engineers to try to fight maintenance cuts at all costs. They are almost always a false economy, as they will lead to much worse longer-term consequences and financial damage for the organisation in the future. Most corporate directors understand the concept of risk management, but unfortunately they don't usually have the knowledge of good maintenance strategies to see the risks they are incurring in this particular context. Using a risk matrix is a useful tool in educating non-technical decision makers.

No matter how sustainable a building is intended to be from design through to construction, it's only when running the building that this potential can truly be realised. Therefore, it's crucial that this is operated and maintained properly. Referring once again to the energy hierarchy, we can see that maintenance reduces the need for energy. 'Green bling' buildings with poor maintenance regimes will never be green. This fact cannot be overemphasised.

So, how do we address this? Throughout the lifecycle of operation and maintenance, we need to ensure that there is a continuous programme of training for occupants, maintenance staff and facilities managers, so that they understand not only sustainable design principles, but also key elements of good operational management for the building and the building services. This need for knowledge doesn't just extend to the building services themselves and how they're actually maintained, but also to the resources used to maintain them, such as the consumables that are used in the maintenance programme. These are similar, in fact, to cleaning products, in that we should be using those with minimal environmental impact. We should also, of course, be using responsibly sourced products and materials. These should be biodegradable, minimally toxic (and, if possible, non-toxic), sustainably sourced and responsibly used so that waste is reduced. We also need to make sure that we are using materials that can be recycled where possible.

Above all, a good maintenance regime should be tied into using the full range of functionality of the building energy management system. If a building has a good building control and management system in place already, it can be well worth looking at the maintenance add-on packages that go with these systems. Integrating these is often beneficial in the long run. There is ultimately a positive link between a well-planned preventative maintenance system and the productivity and well being of both a building and its occupants. (See Chapter 14 for more discussion of behavioural change.)

There's a sequence of diagnostics to done, which are not obvious to building users but which are nonetheless very important, as unless they are done, there is no way of knowing the status and efficiency of operation of the plant and equipment. In this sense, an important dimension of maintenance is risk management.

Because maintenance is behind the scenes, like commissioning, and as such is hidden from view, we tend to only notice it when something doesn't work or looks dirty. Most maintenance is very subtle, such as the carrying out of diagnostics. This should take place using a prescribed regime, often linked in with the manufacturer's requirements to ensure compliance with warranty terms. In fact, many manufacturers give very good maintenance guides and regimes. A good example of this might be a boiler check of combustion efficiency using a combustion gas analyser. Another is the link to air conditioning inspection both for maintenance and now required under the European directive for energy performance (EPBD) for air conditioning energy efficiency certification (Figure 12.5). CIBSE's TM 44 provides the methodology and standards, which includes recommendations to improve the efficiency of air conditioning systems. These might range from improvements to the maintenance regime, through changes to the way the system is actually operated, to providing recommendations on the specification of a new or replacement system. This is a good example of how inspection and maintenance can be enhanced with the connection to energy efficiency and good practice. In my

Figure 12.5 Measuring temperature performance and efficiency

opinion, this should be radically extended to cover a whole range of plant and building equipment. The more this is legislated for, the better it will be for achieving a regime and an incentive to create a more sustainable building, in terms of good practice for building services operation.

Also, maintenance gets a brief mention in the building regulations, primarily to prove that systems comply with the regulations to show that they have been correctly installed and commissioned, 'with the provision of operating and maintenance instructions for users' (Part L 2010). Another way of showing compliance with the regulations is to produce information following the guidance in CIBSE TM 31 Building log book toolkit (CIBSE 2006). All of this is a step in the right direction, but it is unlikely to succeed without proper enforcement which to date has not happened (see legislation, Chapter 3).

Thermal imaging – seeing in a different light

Using thermal imaging as part of a PPM regime is one of the best and most practical uses for this tool. As a regular maintenance and diagnostic tool, using thermal imaging has one very strong advantage in that it is used as a totally non-contact maintenance and measurement too, that allows maintenance engineers to work at a safe distance from moving or hot machinery and electrical infrastructure without the need to switch off or take the building services plant out of service.

All moving equipment can be effectively targeted for monitoring as part of PPM. Generally, long before most plant and equipment fails, there will be a steady and sometimes significant rise in the operating temperature of certain components. This can be, for example, as a mechanical fault develops in motors, compressors, fans, pumps, bearings, shafts, belts, gearboxes and even the casings (Figures 12.6 and 12.7). All components that emit heat can be measured against standard performance benchmarks to highlight anomalies, to indicate the increased possibility of component failure.

Thermal imaging can be crucial in identifying faults in the building's electrical systems, specifically monitoring the physical connections, terminals, components and cables (Figure 12.8). The physical state can then give a guide to any problems that may arise through electrical load imbalance, potential overloads or problems with the harmonics in the electrical system. All this leads to an effective PPM to identify potential fire hazards and possible equipment failure.

The information gained from a thermal imaging inspection can give an excellent guide for establishing prioritised maintenance schedules as it identifies what plant and equipment needs immediate maintenance, and acts as a predictor of which items are more likely to deteriorate and need future repair or prioritised servicing. Performing this type of PPM can therefore significantly reduce or even eliminate the need for more extensive and potentially expensive

Figure 12.6 Pumps and motors

Figure 12.7 Motors and compressors

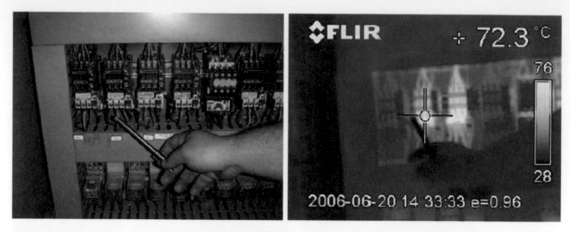

Figure 12.8 Electrical survey – checking circuits and panels

Figure 12.9 Under-floor heating circuit – maintenance and commissioning

Figure 12.10 Very effective and easy checking for chilled ceiling operation

Methodology for using thermal imaging in a PPM regime

Start with a baseline survey and then track changes to the thermal images and recorded temperatures of the plant and equipment over time. Use manufacturers' specifications and guidelines for measuring predicted performance and standards for the operating conditions of your systems and equipment and use these as a reference line.

This will produce a series of baseline images which can be used for comparison with future images taken as part of the PPM schedule. Elevated temperatures and specific hot spots that weren't recorded during previous thermal surveys may indicate developing problems. These regular surveys will then assist in identifying trends, help establish key indicators, and give a comparison for checking the effectiveness after any repairs or maintenance, to indicate if the work was effective and successful.

Integrate the thermal imaging programme with other predictive maintenance methodologies, techniques and programmes. Thermal imaging can be very effectively integrated with a number of other PPM technologies, such as ultrasound surveys, pressure testing, vibration analysis, motor circuit examination and conventional temperature measurements. All of this collected data makes for a very comprehensive and effective PPM programme.

This method can also be very effectively tied in with the commissioning and continuous commissioning process (covered in more detail in the previous chapter). It can be used to survey new plant systems and equipment and to establish a baseline measurement. It can also be used as a verification tool as part of any new equipment acceptance process. I have even seen part of a tender specification make it a contractual obligation for thermal images to be recorded prior to the equipment being delivered and fitted to the plant.

repairs that could arise, and therefore prevent larger-scale plant or system failures, and therefore have a positive effect on future building sustainability as well as increasing safety, reliability and efficiency in process plant production.

Thermal imaging is also a very useful tool for many elements of commissioning, and it also leads to repeated use for continuous commissioning and maintenance procedures. Checking performance and functional parameters such as flow and return temperatures in heating and cooling circuits (Figures 12.9 and 12.10).

Application of thermal imaging to sustainable buildings

Examples of what can be scanned using thermography for buildings and building services engineering:

- buildings – external fabric and building envelope
- flat roofs integrity
- building plant: boilers, chillers, refrigeration

- electrical components and systems: switchgear, breakers, bus connections and contacts
- transformers and associated connections
- mechanical couplings on rotating equipment
- process pipe runs and heat exchangers
- compressor heads and component parts
- motor and generator components, connections, windings, feeders and exciters
- bearings
- drive gears and drive belts (for excessive wear and resultant friction)
- steam systems, steam traps and pipe runs and integrity of insulation
- tank levels and insulation problems

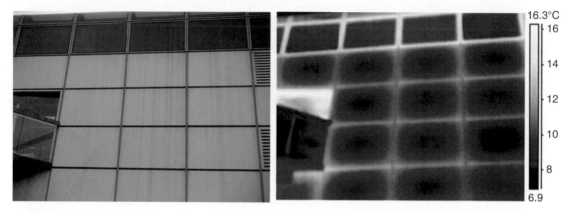

Figure T12.0 External building fabric – leaking like a sieve

Figure T12.1 Survey of building fabric

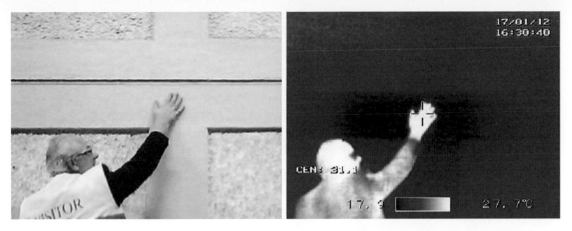

Figure T12.2 Building fabric – cold bridging from uninsulated steel frame

Loss of heat detected, venting from the apex of a roof covering a swimming pool

Figure T12.3 Excessive heat escaping from a building

No taping of joins in insulation board

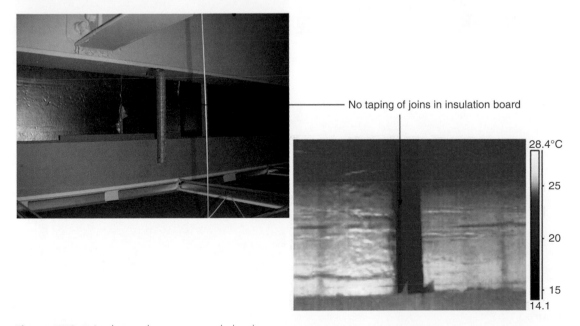

Figure T12.4 Insulation above a suspended ceiling

Poor seals on doors

Figure T12.5 Checking for seals on doors to stop heat loss

Figure T12.6 Behind the plasterboard ceiling reveals chilled ceiling

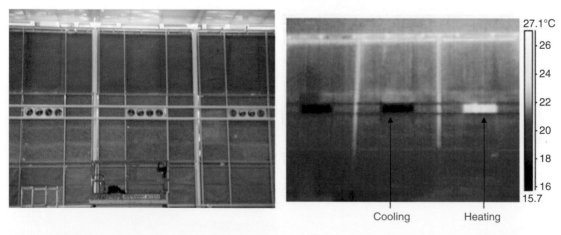

Cooling   Heating

Figure T12.7 Heating and cooling running at the same time – diagnostic to find out problem with sensors

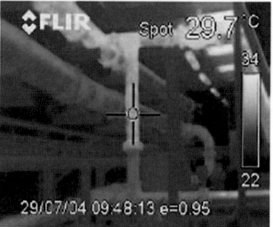

Figure T12.8 Measuring flow and return temperatures for maintenance and commissioning

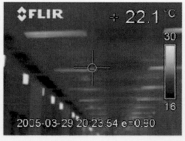
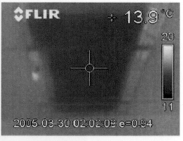

Panels in centre of picture
are not functioning

Passive panels (left): Active panels (right)

An example panel 518 (floor 2)
at too low a temperature

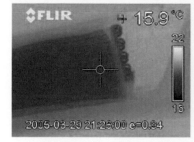

Details of passive panel

Figure T12.9 Detailed diagnostic survey of chilled ceiling operation

Figure T12.10 Panoramic view of swimming pool area

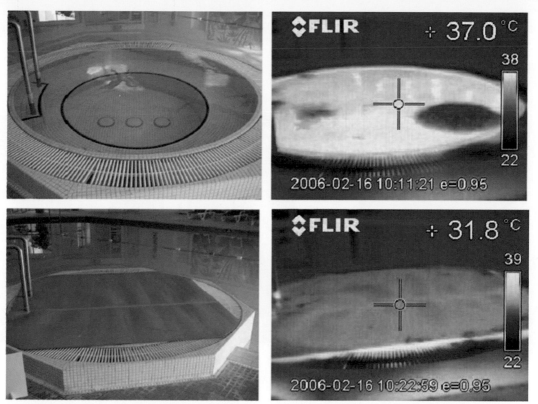

Comparative before and after to show quick effect of covering a spa area ro conserve evaporation and heat. A purpose-made close-fitting cover would show better results.

Figure T12.11 Simple cover to conserve heat and evaporation

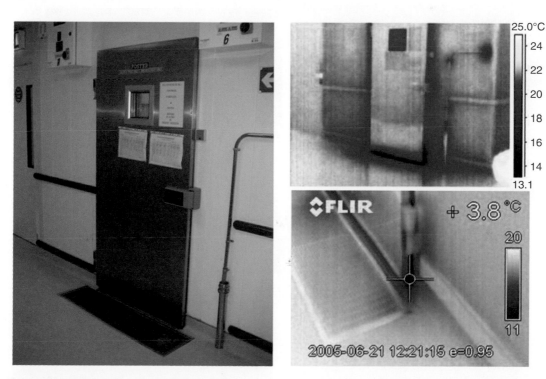

Figure T12.12 Checks on freezer door seals

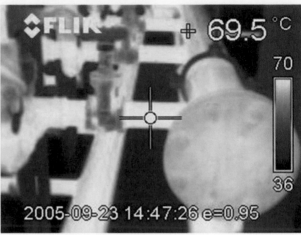

Figure T12.13 Backend of boiler pipework – insulation needed!

Figure T12.14 Boiler plant checks – insulation needed!

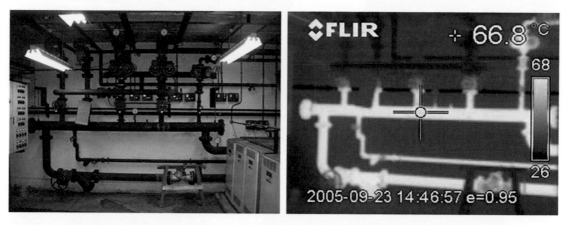

Figure T12.15 Pipework checking – insulation needed!

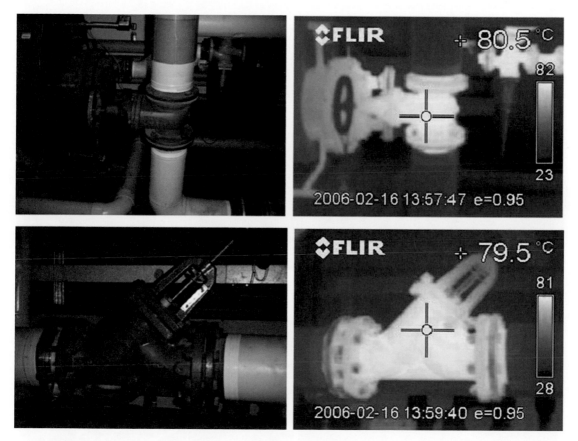

Figure T12.16 Checking valve function and performance of system

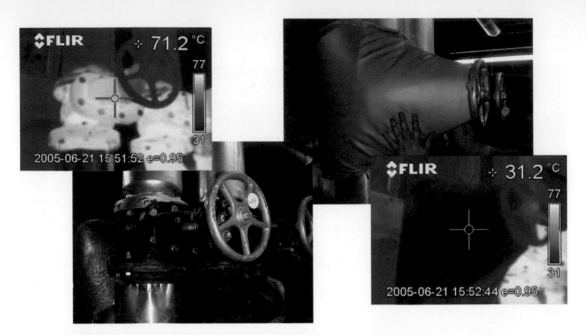

Figure T12.17 Before and after – effective use of Velcro valve jacket proves the value of insulation

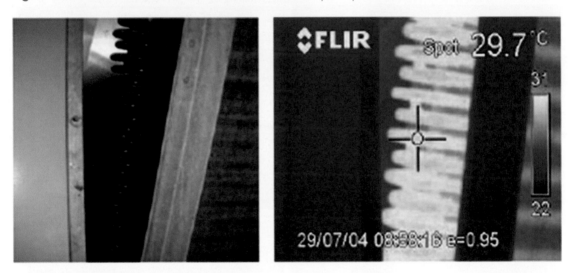

Figure T12.18 Measuring cooling tower pipe coils

Figure T12.19 Heat exchanger uninsulated

Ice is an insulator

Before and after maintenance – shows the impact: –2.1°C to –17°C

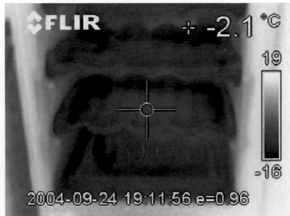

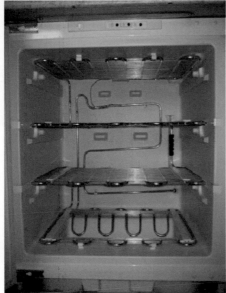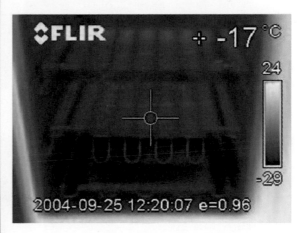

Figure T12.20 Value of defrost maintenance

Linking sustainability and maintenance

There has been a systemic failure in joined-up thinking between sustainability and maintenance. In many buildings I have visited, I have found a well-meaning facilities manager announcing a new green initiative, only to find that the campaign ignores the central elements of maintenance. So we need to change our culture – once again, turn to Chapter 14 on behavioural change for more on this.

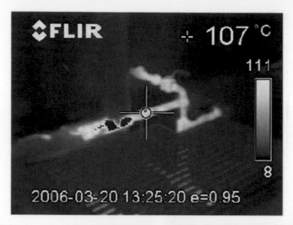

Figure T12.21 Measuring very hot and very cold plant and services

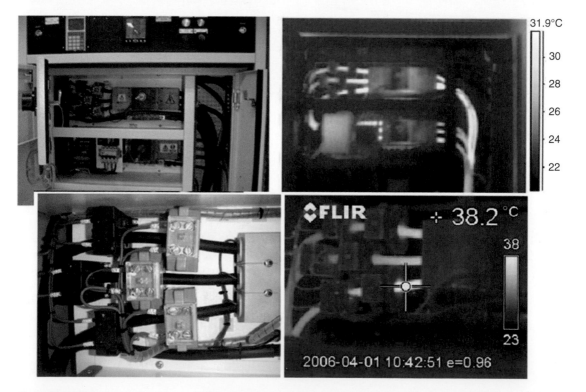

Figure T12.22 Electrical survey – checking circuits

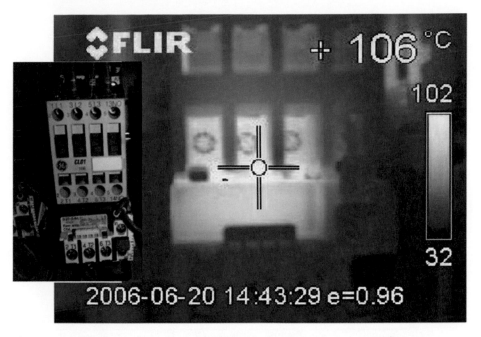

Figure T12.23 Electrical inspection – high temperature detected on a extract fan contactor

I cannot stress enough the point that to implement a sustainable agenda successfully, we need to take a holistic view. After all, there are a lot of similarities between all the mechanisms we can use throughout this book. Maintenance is inextricably linked to the wider sustainability agenda. So how do we make maintenance part of sustainable engineering?

For a start, we need to broaden the standard '3 Rs' often associated with sustainability: reduce, reuse, recycle. We need to add 'repair' to this list, because maintenance is integral to the efficient reuse of most components in building services. These 3 (or 4) Rs are the waste hierarchy (Figure 12.11), and we can easily see the link to the energy hierarchy discussed in the Introduction and referenced throughout this book. In fact there is a fifth 'R' – Recovery where, after the previous stages, it is possible to recover energy by burning in a combined incinerator, heat and power station. This is the last resort before disposal or landfill.

Overall in our culture, maintenance has suffered with the rise of the throwaway society. There is always a dilemma between the cost of goods and the cost of the time and materials needed for successful repair. Short-term economics is working against sustainability. Too often, it's cheaper to throw something away, be it a phone or a car, than it is to repair it. In other parts of the world, however, this is not the case. In India, for example, society abounds with the kind of resourcefulness that underpins good maintenance. Walking down the back streets of Delhi, I witnessed almost every conceivable intervention to keep aging consumer goods running, which demonstrates that it can be done. In the West, on the other hand, things are too easily consigned to the scrapheap as being 'beyond economic repair'. Unfortunately,

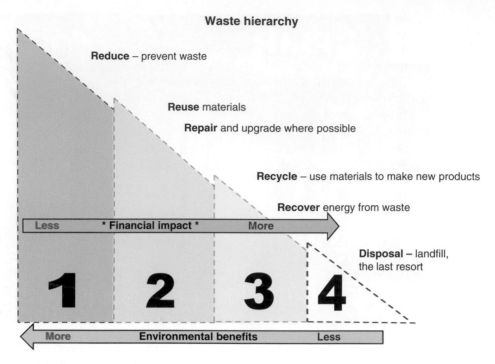

Figure 12.11 The waste hierachy

when we make the decision to throw something away on these grounds, we are not looking at the hidden costs that go with that decision. In reality, there is the cost of disposal and also the cost to the environment, not to mention environmental taxes. Too often, these are not taken into account. I would also argue that there should be lower VAT rates on repairs, just as there are on insulation and other energy-saving products. By lowering VAT on repairs and maintenance, there would be a financial incentive to choose this option. This ties in with the encouragement of sustainability overall.

The development of user-friendly operation and maintenance (O&M) manuals is extremely important. These need to be integrated with the range of building services functions and tied in with an asset management structure for all building services. The most important part is to make sure that this links in with the building log book so that everything is brought together – see Chapter 11, where this forms part of the discussion on commissioning.

This also links with the contractor's role. It's a vital component of the contractor's work, that they take full advantage of the business opportunities of offering a comprehensive maintenance service. This is a role that the building services engineers on site and the contractors need to emphasise to facilities managers and building owners: the importance of a good maintenance regime.

Indeed, at the time of writing, contractors are working in a shrinking and difficult market, so they should be looking at becoming experts in the whole life of a building, and offering a portfolio of services including maintenance.

There is also another role for maintenance contractors to check energy efficiency, because in practice they have already doing 50% of the work without thinking about this. Planned preventative maintenance fits in perfectly with energy auditing – they're already collecting the data in the course of their work!

Ultimately, all these small but important items lead to excess energy use and loss of performance, which compound to form a counter to what was part of the original intent – a sustainable low-energy building. This example gives us the crucial role that planned preventative maintenance fulfils in the sustainable building process. Operation and maintenance is the final, crucial part of the process outlined throughout this book. I use the word 'crucial' advisedly – in terms of the total cost of ownership, the operation and maintenance of a building accounts for 60–85% of the total lifecycle costs (NIBS 2010).

It's clear that a sustainable building must operate to optimum performance if it is to function correctly, be energy efficient and create an optimum adjusted and beneficial indoor climate for its occupants. Any shortcuts or cutbacks to maintenance fly in the face of sustainability.

The need for standards and legislation to underpin and raise the importance of maintenance has never been more important than now in making the links for the core issues of maintenance and energy efficiency and its key role in delivering a continuous programme for delivering sustainable buildings.

References

CIBSE (2006) TM 31 *Building Log Book Toolkit*, The Chartered Institution of Building Services Engineers, London

CIBSE (2007) TM 44 *Inspection of air conditioning systems*, The Chartered Institution of Building Services Engineers, London

Part L (2010) Conservation of Fuel and Power, Approved Document L2A, HM Government

NIBS (2010) *Construction Guide for Specifiers*, Section 01 81 10 (01120) Facility Service Life Requirements, National Institute of Building Sciences

RICS (1990) *Planned Building Maintenance: A Guidance Note*, RICS, London

3   The human element

13 The skills challenge

You never stop learning in the building services industry. It's a practically based profession and so all of us are continually adapting, innovating and, of course, finding solutions to problems as they inevitably occur on site and in the wider sphere of contracting. Practice does, of course, need to be underpinned by technical knowledge and theory, but you just can't learn all there is to know from a book. Therefore, it's essential to carry out continuous training and development. I've always maintained that a skilled workforce will lead inevitably to higher standards and will make companies more profitable. This is because all the associated up-skilling and best practices will make it hard for the cowboys to survive in our industry and, thus avoid the knock-on effects of some poor practice we see from time to time.

There used to be a predominance of traditional apprenticeships in the industry, but they have diminished over time. This is a matter for deep regret, because it has eroded the skills base. We could face a generational gap in skills as a result of this short-sighted practice; this is a real challenge for the industry. As the economy recovers from recession, there will be many areas where there will be a severe lack of skilled, experienced people for different tasks and activities. In fact, it is an ideal opportunity to consolidate and up-skill the workforce while the economy is slack, so that the industry is prepared and ready for the challenges ahead.

Resurgence of skills

Ultimately, I believe there *will* be a resurgence of skills; this will be driven by the development of the market for low carbon and renewable technologies that will grow significantly in the building services sector. These are considered new skills, which will bring an emphasis on training, as the industry changes from traditional carbon-based technologies to lower carbon and renewable energy. This will create many opportunities for skills enhancement and retraining. In my view, this will be a catalyst to bolster the industry although it doesn't solve the problem of replacing the quality of training that was once obtained via traditional apprenticeships over five years and more.

This issue needs to be put on the agenda. The skills agenda needs to encompass industry standards and drive mechanical and electrical training

Delivering Sustainable Buildings: an industry insider's view, First Edition. Mike Malina.
© 2013 Mike Malina. Published 2013 by Blackwell Publishing Ltd.

schemes as well as short-term up-skilling courses to firstly establish and then maintain a competent and sustainable workforce. At the moment, people can study how to install photovoltaic panels in a week. Is this satisfactory? I would say not. Providers challenged about this are saying that their courses are for experienced people in the industry but I'm not sure that anyone can retrain in such a short period of time.

That said, I think that it's a mistake to view all new approaches in the industry as totally 'new' technologies. Many of them, I would argue, are not new as such. Rather, they are just a redevelopment or adjustment of existing industry practice. For example, if you're fitting solar hot water, all you're doing differently from the old approaches is changing the heat source. All the traditional skills are still needed, as if you were fitting a conventional boiler. You still need to understand pressurised systems, pipework and the fitting of pumps, valves and expansion vessels. One additional skill that may be new for some is safe working at heights – much of this technology will be working on roofs. There are also opportunities for working in partnership with other trades, in this example there is the potential for electricians to work on joint projects with roofers. So the potential for business diversification and collaboration is a benefit in the developing market.

As this market develops, we will need a programme of enhanced retraining and upgrading of skills. The key will be integration of technologies, sometimes known as systems integration – see Chapter 10 for more information on this. Otherwise, there is a danger of history repeating itself, in that certain parts of the industry were pigeonholed in the past, with different technologies designed and installed by specialist building services engineers. Now, we all need to become generic engineers, because we're integrating the technologies so that they become truly electromechanical and controls based.

The current demarcation is crazy – waiting for electricians, or other specialist engineers, is terminally inefficient, outside the more involved and highly specialist areas. Therefore, it's absolutely key that the skills challenge is coordinated, and that standards are benchmarked and validated for a proper programme with quality standards.

Construction and refurbishment of buildings has lots of different elements, so we need joined-up thinking in order to get a holistic view. For it to work efficiently, it must be all joined up. Unless all the different functions and types of technologies are coordinated properly, we're going to get poorly functioning buildings. Despite this, however, too often people see only their small part of the process and rush though their work, already thinking about the next job; this doesn't give proper value and could endanger the quality and effective completion of the job.

Case study – East of England skills challenge

This is developed from my own experience as the Eastern Regional Chair of SummitSkills.

SummitSkills – the sector skills council

A very positive development is the work of SummitSkills, who are the sector skills council that is tasked with developing building services engineering skills. An excellent interactive careers map has been developed to help and guide a range of possible paths at all levels for the industry (SummitSkills 2011a).

SummitSkills is an employer-led body, which lists the following as its six key objectives:

- employer engagement
- offering expertise, safeguarding standards
- enhancing quality and delivery
- raising ambition
- effectiveness and evolution
- partnership approach

It is involved with the whole range of vocational qualifications – from preliminary diplomas to degrees – and it also offers a training standard for trainers. Personally, I've always believed that to train in this industry you should be a licensed trainer. There should be a global standard, just like teachers have to have universal teaching qualifications. At present, people can go to any private trainers. Do they know what they are getting? Is it accredited, and valid for what they are trying to achieve?

SummitSkills also carries out a large amount of research within the industry. Recent work includes a recession impact study, and an apprentice cost–benefit analysis. This latter research found that employing apprentices could save a company up to 15% in labour costs on a large project.

The government has awarded several million pounds to SummitSkills to develop a National Skills Academy for Environmental Technologies (NSAET 2011). This will be a network of accredited training providers, mainly colleges of further and higher education, who will be coordinated and brought together under the banner of this national skills academy. The government is only providing some of the funding for this – the rest is expected to come from employers – but, perhaps because of these constraints, the academy has been organised in what seems to be an imaginative and efficient way, with a hub and spoke arrangement of training providers. In this way, it can deliver its vision of a national centre for green skills development, helping to ensure that we have the skills to deliver a low carbon economy in the future. This will make the sector more competitive, as well as greener, and will help to ensure the survival of the sector into the future.

National Skills Academy for Environmental Technologies, NSAET

The Skills Academy is in place to raise the quality of training and ensure that only industry recognised competency-based training is delivered from craft to professional, from spectator to practitioner. It is already working with employers and partners to ensure that the sector workforce has the right skills, at the right level, to enable them to be competent, productive, effective and efficient.

This kind of central body is exactly what's needed to create uniformly high standards. The key skills covered by the academy will be vital for the transition to a low carbon economy – such as quality specification, design, installation, commissioning and planned preventative and reactive maintenance. What will be essential will be integrating these key skills into the development of renewable and low-carbon technologies.

Other drivers for the new low carbon technology skills agenda will be feed-in tariffs (FITs), the Renewable Heat Incentive (RHI) and the need to become members of a micro-generation certification scheme. These will ensure that training provided by the NSAET and its network will be essential. Contractors won't be able to access the financial incentives such as FITs and RHI without it, so they will have to have the training to become members of those schemes. This is an example of how almost mandatory training makes business sense to get access to the scheme incentives and funding no matter what, because it enhances the workforce.

This will include all the low carbon and renewable technologies such as:

- heat pumps
- mechanical heat recovery and vent systems
- combined heat and power
- the range of biofuels and biomass
- micro wind energy generation
- micro hydro generation systems
- solar thermal hot water
- solar photovoltaics
- rainwater harvesting.

This is not a complete list, but gives a flavour of the enhanced skills that will be needed for the industry. Some of this knowledge can be simply up-skilled from existing competencies that are already taught under various training programmes such National Vocational Qualifications (NVQs).

One very important development is the launch of the Green Deal Skills Alliance (GDSA), which will create new training and accreditation opportunities for the energy assessment, advice and installation workforce. It will be made up of three sector skills councils including AssetSkills, ConstructionSkills and SummitSkills.

Remember, we won't just need to master the technologies themselves; we will have to include the issues discussed in the Introduction and Chapter 5 about lifecycle costing, energy flow, reducing waste, maximising water efficiency and seeing the wider picture. This awareness will be crucial to all training and awareness as we move forward in the developing low carbon economy, making both the links on the technology front and the vital social changes that will be needed in the move towards greater efficiency From a social economic perspective, it will mean reducing fuel poverty (this includes everyone who has to spend more than 10% of their income on fuel, which leaves people scared to turn heating on) and carbon emissions. Even if we didn't believe in climate change, it undeniably makes good economic sense to save resources, energy and money!

Manufacturers and trade associations – role in training and skills

We need to plan for the future by building on good practice from the past, such as the work developed by quality training providers and those manufacturers that do excellent training on products. Although manufacturers will obviously deliver product-specific training, this can often be used to enhance skills generically, due to the similarities between products. Continuing professional development (CPD) plays an important part in connecting the industry with manufacturers and keeping engineers and support staff up to date. I am often involved in giving an independent view and overview with this type of training (Figure 13.1).

SummitSkills Manufacture and Sustainability Group

This group was established to ensure that we link the development of technology to present and future skills requirements, providing a knowledge

Figure 13.1 CPD training is important to the industry

platform to engage in, consult with and develop new and innovative technical applications and products. It also ensures that the process between production and delivery into the market includes development of the skills required for the workforce.

Trade associations

There are also good, high-quality programmes from trade bodies, such as B&ES. This association of leading high-quality contractors also licenses accredited trainers – I'm one of them and have trained over 500 contracting engineers, so I can see from experience the real value that these programmes deliver. On the electrical side, the ECA also delivers very high quality training. The professional institute representing building services engineers, CIBSE, also runs a variety of courses, including CPD on a range of specialist areas.

As well as needing skills on the ground, practitioners need to keep up with legislation, standards and current awareness. Legislation changes regularly, and people need to be aware of this, since they have to comply with it. For example, the building regulations are now scheduled to change every three years. One answer to this is the B&ES's Building Engineering Services Competence Accreditation(BESCA) This allows contractors to become a 'competent person company'.

BESCA competent persons scheme

Accredited contractors are able to self-certify without needing to apply for prior approval through their local authority for compliance with the building regulations for the installation of mechanical and some electrical building engineering services.

To ensure a high degree of compliance with the building regulations, the Department for Communities and Local Government developed the competent persons schemes that allow suitably qualified firms to self-certify any work that they carry out, which is covered by the wider scope of the building regulations. This avoids the need for inspection by local authority building control departments.

The Building Engineering Services Competence Accreditation Ltd (BESCA) was established by the HVCA (now the B&ES) to develop and operate a competent persons scheme in relation to both Part L1 and Part L2, and to Parts F, G and J, which relate to the installation of the full range of fixed mechanical and electrical building services.

Companies that qualify undergo an independent third-party quality and systems inspection assessment and audit. The employees who administer the certification also have to demonstrate an appropriate level of qualification and competence.

The green deal – implication on skills

The UK coalition government, elected in 2010, introduced an energy bill that makes provision for a 'green deal', the terminology seemingly borrowed from Barack Obama's green jobs initiative. Whilst this would seem to be a welcome development, there is a fear that the government hasn't worked out the scale of what's involved or the number of people needed to deliver it. The Energy and Climate Change Department announced in a press release that there would be 1000 green deal apprenticeships, but this is a drop in the ocean considering that it needs to support 100,000 workers by 2015, with the potential to grow to 250,000 at the peak when the low carbon economy is fully established. It seems that the government doesn't comprehend the scale of the undertaking. It doesn't help that the responsibilities for implementing this are split between different government departments – Business, Education, Energy and Climate Change, Communities and Local Government and the Treasury all need to be involved – it's akin to a confused octopus! For this to work as a coordinated political and economic measure, it needs joined-up government, because the skills and training need to be universal in terms of standards. Quality is vital. We don't want to see regional disparities, or the problems in quality that have happened in the past with schemes such as the Youth Opportunities Programme.

On the other hand, one positive development is the proposed coordination of the six major built-environment skills councils, including SummitSkills and ConstructionSkills. There seems to be a possibility of a Built Environment Skills Alliance, with the idea of formalising the links between construction and property. This is being done with a view to ensuring standards for things like insulation (cavity wall, loft , external and internal) through to hopefully more integration between other built environment skills and the building services industry.

It is to be hoped that the green deal approach and its funding will involve development of training, based around the energy hierarchy, utilising all the skills of the different disciplines to reduce the amount of energy being used. It's all about getting control and managing energy more effectively, and then finding ways of generating heat and power from low carbon and renewable technologies. This will need coordination, and also a complete integration and update of national occupational standards, which form the basis of educational qualifications and NVQs. This should be linked in with industry skills cards for all operatives in the industry, which will then guarantee certain standards being met. As the former Secretary of State for Energy and Climate Change, Chris Huhne, has stated, there is no place for cowboys in the green deal.

For this to happen, the government must give backing to all the sector skills councils to ensure that the quality standards are maintained. To cite a warning from history, as I mentioned in Chapter 1, we don't want to create the equivalent of the double glazing industry of the 1970s and 1980s. The

Figure 13.2 Green Van Man – © Sarah Malina

challenge, especially in the domestic market, is to educate both the consumer and the industry as a whole before the low carbon market develops beyond expectations. There have already been cases of exaggerated claims on performance and payback of some technologies as well as inappropriate installation and siting (Figure 13.2). This is also covered in Chapter 1, where we look at rigorous standards and enforcement.

The biggest challenge, of course, will be finance. The green deal starts at the end of 2012. Finance from the utilities and private companies will be made available to fund these projects, so that the householders don't have upfront costs. The costs of the upgrade work will be paid back by savings on energy bills, so householders will carry on paying what they would have paid despite their now lower-cost heating to pay back the loan. It's a clever idea, but it has its negatives – there will have to be vigorously audited national measuring to check that savings are indeed being made, because habits around energy use have to be factored in.

For example, domestic energy use is currently going up because of the introduction of more gadgets and remote controls. Some of this technology will change with the implementation of EU Directive 2005/32/EC, which deals with the eco-design of standby and off modes in electrical equipment. This means that eventually all new devices will have to have a built-in ability to be powered off completely. At the moment, plugs and chargers constitute so much wastage. However, this has to be integrated with more efficient appliances to reduce energy consumption overall.

Considering current habits and energy use, we can see the importance of measuring what is happening. Under the green deal proposals, the idea is that the measures that are applied to the house and not to the person. Therefore, if you move, the new inhabitants of your old house inherit the energy project. But what if one person leaves and a family of four move in, or the other way around? Even without changes in the number of occupants, it might be that one person might be really energy efficient, while the new occupant leaves everything switched on.

The green deal is a step in the right direction, but it will need a coordinated and well thought out programme if it's to make any impact on carbon reduction. It will also create the potential for other new jobs and skills, because we're already seeing the concept of a green deal advisor, who will need training to at least NVQ level 3. Utilities companies will undoubtedly pick such opportunities up – the only question is does this create something of a poacher turned gamekeeper situation? It does seem somewhat strange if we're asking utilities companies to lead on saving energy. They have, in the past, scattergunned some ill-thought-out energy saving schemes, such as issuing compact fluorescent lights for all. These were distributed to thousands of homes without anyone asking questions about the fitting type the householders were using. This meant that at least 50% of the lamps were left in people's drawers because they didn't fit, or the utilities companies hadn't explained how to use them properly. It became a somewhat counterproductive tick-box exercise.

On the building services engineering front, there will be a need for more heating and air conditioning fitters, to fit the range of gas condensing boilers and multi-fuel arrangements and the potentially large number of air-source heat pumps. Manufacturers will also be key in the development of processes providing significant amount of product training, and their links with sector skills councils need to be strengthened.

Knowledge platform

One of the key strands that is related to the green deal but does have a more general application is the communication of knowledge and understanding to customers and clients. The need exists to ensure that customers understand what 'low carbon' technology is about and better explain the benefits without introducing a 'black art effect' and shrouding it in mystery. The fact remains that the more you understand the more likely you are to open your mind to the options.

SummitSkills, AssetSkills and ConstructionSkills are the main bodies in the sector. They are already looking at designing new apprenticeship frameworks, and updating and revising existing ones. New entrants can be trained for the low carbon industry, existing engineers can be retrained for the industry as the sector grows. Kingfisher Group, the retailer which owns B&Q, is now running retail City and Guilds courses for their staff, so that they can acquire the skills needed to sell greener products. This is very important, since the wider public need to be educated to make the right choices. An informed retail sector will certainly help, although of course it needs to be impartial. Hopefully, this Kingfisher initiative will become an industry standard for all retailers involved with these products. It would be good to see a complete up-skilling of the standards in the supply chain, from manufacturer to wholesaler or retailer to installer.

SummitSkills is working with employers and their representative organisations to develop a Green Deal Competency Framework. This will be an integrated portfolio of national occupational standards (NOS) and qualifications to identify the standards of work and knowledge required to become an energy assessor, advisor or installer.

SummitSkills has published its environmental strategy document through to 2013. This document describes how SummitSkills will be working on behalf of the industry to promote and gain a range of opportunities to help the UK move towards developing the skills to achieve a low carbon future (SummitSkills 2011b).

One of the biggest apprenticeships schemes will be developed by British Gas, a major player in the process. In the past few years they have created over 1000 new 'green collar' jobs. They are also planning to recruit another 2000 people as the process develops. British gas obviously sees business development here.

There will also be a role for local authorities and government to assist and develop partnerships for industry, local business and the local population through the newly established local economic partnerships (LEPs). This will be vital for encouraging a more joined-up approach in creating the programme of renewal as we move towards a low carbon economy.

Ultimately, of course, we also need a raising of awareness for the end user. For this to work, it won't be a 'supply, fit and forget' process: it's got to be a continual up-skilling process, encompassing awareness and education for the entire populace. It will be about winning trust and support across the supply chain through to the installation and use by the end user. History has shown that all through technological development people have become more aware and, within a comparatively short period, they have adopted the technology until it seems to become the norm. Looking at recent times we only have to look at how satellite TV, mobile phones and the internet have impacted society. This I believe will be the same for low carbon technologies as they develop with economies of scale and production. Most people will start to see the benefits as fuel prices continue to rise. Hopefully, people will make the links between saving money and just being more environmentally aware and resource conscience.

It would be great if the energy hierarchy could be made a popular method for achieving this. From school to wider information campaigns, the government and indeed the industry need to do a lot more.

References

(NSAET 2011) The National Skills Academy for Environmental Technologies www.nsaet.org.uk

SummitSkills (2011a) http://www.summitskills.org.uk/careers/343 (accessed 21.8.2012)

SummitSkills (2011b) Strategy for environmental technologies in the Building Services engineering Sector 2010–13: http://www.summitskills.org.uk/public/cms/File/Renewables/SummitSkills%20environmental%20strategy.pdf (accessed 21.8.2012)

14 Changing behaviours

However much attention we pay to technology and processes, they cannot, on their own, improve the viability of our energy needs or our attitudes and understanding of our own building environment. Any change in our consumption habits or behaviour in the built environment has to involve the people who actually use the buildings.

Behaviour, attitudes and perceptions

There's a whole range of factors that govern people's behaviour and attitudes. If a building is not functioning correctly, if it is badly commissioned and provides a poor environment, people will complain. This is human nature. If people feel unhappy, there can be a spiral of discontent. This affects both productivity and industrial relations. Once people have formed a negative view of their environment, it can be difficult to regain the ground and win people round to seeing it positively. Facilities managers will often find it tempting to take easy options in the hope of improving people's views, for example by buying fans to improve people's perception of the internal air temperature during hot weather. This provides a degree of perceived cooling, but the correct solution might be better use of natural ventilation. However, in order for more effective and sustainable solutions to be successful, the facilities manager would need to attend to the education of occupants. Education is vital to the process.

Tackling workplace behaviour is a major issue that needs to be addressed. Without this, the whole process of creating sustainable buildings will miss a major part of the armoury in its potential success. Educating people is vital to what we're trying to do. This element has been really neglected to date. I find this very frustrating. Perhaps it hasn't been covered enough in the media. If you read about something all the time, you will be educated. Unfortunately, the only time I see energy mentioned, it's invariably either about bills rising or about the fact that we might run out of energy altogether – although this in itself should be enough to get people moving and feeling concerned! But what we are missing is the information about how people can help the situation themselves.

Delivering Sustainable Buildings: an industry insider's view, First Edition. Mike Malina.
© 2013 Mike Malina. Published 2013 by Blackwell Publishing Ltd.

Let us consider what motivates people. There is quite an interesting change between people's behaviour at home and at work. Psychology has some of the answers as to why.

There have been some interesting studies – a research project looking at the domestic side of energy use looked at the motivational effect of setting targets for two groups of households. The two groups had two different goals. Of the 80 families who were asked to set a goal to reduce energy consumption during the summer, half were asked to reduce consumption by 20% and the other half by 2%. The researchers gave the families feedback on how they were doing three times a week. They also had another group of families as a control group. It was found that those asked for a 20% improvement conserved the most, between 13% and 25%. The ones that were asked to do 2% hardly made any change because it was too easy. This shows that if you set a higher goal, people do better and also that ongoing feedback is an important part of the process – people are motivated by challenges and by being able to see their progress towards the goal (Becker 1978). Having said this, the goal or target has to be realistic and wholly practicable.

People often behave differently at work due to several factors. Firstly, this is a different environment where they don't pay the bills. However, there could be good links made here if we want to motivate a workforce. If we can't appeal to their better judgement by talking about climate change, perhaps it would have more sway if we pointed out that by saving energy at work, just as we might at home, we will be saving money which ultimately all stacks up to better profitability and secures their employment. This efficiency argument may be used by some companies, but strangely it is more often used with peripheral costs such as stationery consumption rather than energy. From an industrial and commercial point of view, people are geared up to be more efficient on process and production, but not regarding service needs or their working or office environment. This has been my observation in the many different workplaces where I've conducted energy and commissioning audits.

Another factor that makes people behave differently at work is the fact that they're in a shared environment with far less control than in their own house. This will be due to a number of factors, such as the set-up of the building, or perhaps the occupants have never been shown how to manage their own building environment. The solution could be as simple as showing them the controls available or to have them explained in a company training session. It seems strange that this often does not happen, as compared to the training and effort which companies expend on health and safety. In the health and safety field, risk assessments are common, and there has been considerable change in practice due to developments such as legislation and growing awareness. We should also put emphasis on making people think about how they are using their building environment and energy use in the same way. This should be integrated with all management procedures, from induction to wider company procedures.

In practice, however, energy use is not actively embedded in company policy and practice, particularly in office environments – the only example that is repeated without real commitment or follow-up is the common exhortations to turn equipment or lighting off at night or when leaving the area. This, on its own, doesn't address efficient energy use. Computers are left on all day, when we wouldn't leave them on at home. We certainly seem to have become a fast-paced society, where everything is expected quickly. Still, a simple bit of patience would give us opportunity to use energy far more efficiently. If we are going to a meeting, how long does it really take to put a computer into hibernation? Similarly, we should for example, turn any comfort cooling or air-conditioning off as well as any other non-essential equipment, such as printers or photocopiers.

This disengagement with energy use is not endemic in all workplaces, and factory environments are often more proactive in their approach. In a factory, conveyor belts and machinery are all routinely turned off – it's factored in for energy saving and also for maintenance, as achieving less wear and tear is a big issue. This shows that we behave differently in different workplaces and to get *every* workplace performing better, it just needs some management intervention and planning. I believe that energy saving and awareness should be integrated into the culture in a way that we take for granted when it comes to health and safety and wider management and employment practices. For this to happen will require a change in mind-set, but it's not impossible. There are precedents, as we realise if we think back to the days before risk assessments and health and safety. Look at how this area has developed over the past decade.

Convenience and resistance to change

Convenience has made change more difficult. One widespread example is the way supermarkets have made it a habit that people always expect an open cabinet for refrigerated or frozen produce. What's wrong with opening a door? Have you ever seen a domestic or even a catering establishment fridge/ freezer without a door? It's ludicrous that the store is heated only for the 'wasted' energy used in refrigeration to conflict with this in an endless cycle. One senior supermarket director admitted that their policy on door-less cold cabinets was a 'double-digit contributor to their energy costs – that is more than 10%, a lot more. And doors are not expensive.' Yet even though they know and admit this, change is still very slow.

The supermarkets also blame the government. One executive said 'If the government told us to do it, we could put doors on all our fridges tomorrow. And if all the big chains did it, we would not have to fear losing customers to our rivals' (Guardian 2009). Some of the smaller chains have adopted more closed cabinets, but the larger chains who have experimented with doors at some of their new 'eco-stores' came up against customer resistance and therefore delayed any more deployments.

Another trait that has developed in the retail sector is an open front entrance policy with a warm air curtain blasting warm air into the outside environment. This again conveys excessive waste over convenience. Most people except this without question. 'We can all make an effective difference as we shop, both as consumers and advocates with this common-sense approach to cutting energy waste' (NFWI 2011).

It is right that we should focus on these examples as almost everyone is familiar with these examples. So the solution would be for every supermarket and retail store to close their doors. This needs to be explained and would become the norm without too much inconvenience to the masses. This is the type of action that is a win-win situation for saving energy and changing people's behaviour, attitudes and perceptions. It would also form an example of mass participation to make the links directly between reducing energy and waste. The big carrot in the marketing of this change would be for the retail industry to pass on the cost savings in reduced energy bills to the consumer by reducing the retail prices of their goods.

Getting it right from the start

Where a business is starting from scratch, or moving into a new building, it is easier to achieve working practices that promote sustainability. The key to this is careful handover. Customs and practices are new, so it's an ideal opportunity to start people thinking about how they should be using and saving energy and considering how to use their working environment effectively, as part of a fresh programme of work. In my view, this should be a compulsory part of the handover process. There's some provision in the building regulations, but it is focused only on commissioning and handover of plant and equipment. Furthermore, in health and safety legislation there is nothing that says you have to do this. I think there needs to be some legislation on this, so that buildings are run and occupied appropriately from the start. I would even argue that this process for new buildings and major refurbishments should be integrated as part of the local authority planning process; making it part of planning conditions.

Existing buildings

Existing buildings can be a lot harder to change. You have already got an existing workforce with established bad habits, as well as technologies that may not be so easy to change. Nevertheless, we still have to find a way of establishing a programme and motivating people. As already stated, the key is education. For an established workplace this will involve training programmes which integrate lessons about the importance of building energy and facilities management, with messages empowering people to become involved in the processes of saving energy and to be aware of the quality of their own internal environment. We saw in the case study that people can be motivated by challenges, and linking this with profitability, job security and the potential for workers to benefit financially from savings made, could be the way forward. It's also important to be honest with people. In a hot summer, they shouldn't expect a perfect working environment. You can satisfy people short term with fans – you've made an attempt, and it's low capital expenditure – but ultimately it doesn't actually cool the air. More emphasis will need to be placed on common sense, and convincing people of the practical realities in order to achieve change.

Can technology help?

First, we need a way of understanding both how occupants behave in a building and how working practices impact on energy consumption. There is a lot of technological development in the field of intelligent buildings with enhanced controls (see Chapter 10), but we need to integrate behavioural change for a truly intelligent building with more enlightened and intelligent behaviour. This has to be part of integration with any energy management programme. There has been a lot of work on complex algorithms using environmental sensor based modelling within building energy management systems. There are controls to predict behaviour within intelligent building, such as the smart lift, which analyses people's movement behaviour and predicts when people will be travelling and sends the lift to the right floor to maximise efficiency. Therefore, they don't have to wait so long for the lift and it also uses less energy. In the Chapter 10, I mentioned the use of sensors to create intelligent buildings, tackling issues such as lighting via motion sensors, as well as sensing carbon dioxide levels, temperature and humidity. This technology, combined with educating people, will create an even more efficient and intelligent building. But this can only happen if people are made aware and become motivated, even a little interested in their own space and environment.

In the future, we will see smart metering. But I worry that this will be done in isolation, without looking at the wider issues of building energy use and patterns of activity or explaining how people can utilise and gain information beyond energy bills. The smart meter will give information to the utility supplying company and they'll give you graphs on quarterly or monthly bills, but not live information, unless this is specifically engineered. We need to associate measuring with day-to-day activity. One of the biggest benefits we could see would be live displays that shows energy at that exact moment of live use. No one sees the energy meter in the workplace. Imagine the display being on the finance or managing director's desk; it would be like using the petrol pump, where you see pound notes rolling before your eyes. My recommendation is for every managing and financial director to have a live energy display on their desk. That would help them keep an eye on energy use and become part of a monitoring and targeting approach.

Getting the workforce on board

One of the best methods for getting people's approval or buy-in is to involve them in occupancy surveys, where they will feed back their perceptions and observations to create some benchmark and statistical information on how the building is performing. This can also show where problems can be targeted to improve the overall effect. This can challenge and change perceptions, and should be a continuous process. It will keep the facilities manager alert to changes that have occurred. This can be linked with the process of soft landings, continuous commissioning and a planned preventative maintenance regime, covered in Chapters 8, 11 and 12 respectively. By adopting this integrated approach the chances of delivering a continuously optimised low carbon use building will be greatly increased.

For accountability and wider information, I have seen some companies display their energy use in the reception of the building. This may be considered by some as a gimmick, but it does have a use in publicising the live and historic energy information as well as providing an opportunity to further a programme of motivational change. Ultimately we need top management buy-in to educate people on the 'shop floor', and ultimately the two sides have to get together to manage energy effectively and produce buildings that work and are ultimately working towards the goal of being sustainable.

One pioneering method is the energy management toolkit, utilising energy management through people rather than just the conventional elements of behavioural change. This has been developed by James Brittain from the Discovery Mill, www.thediscoverymill.co.uk. The focus is to help put *people* at the heart of energy management success. This focuses on action rather than theory. It focuses on actively working with local leaders and teams to identify the very best opportunities and to deliver energy efficiency with enthusiasm and success. I certainly endorse this approach and have adopted this method in all the training that I have undertaken on energy management and wider low carbon initiatives.

A lot more work needs to be done on what can be saved through good housekeeping, in order to explain to people that to change habits doesn't cost anything. It's important to analyse barriers to behavioural change, to understand why inefficient behaviours become habits. It's also important to be honest with people. This, in my experience, tends to work with some patience and application clearly planned and thought through. It's also vital to keep messages simple. A grand campaign can go over the top and give too much information, or, even more commonly, tell people of the importance of saving energy or even of 'saving the planet' without giving any practical advice on how to do it. Training should be integrated with targets: there should be a clear set of goals.

People need to be educated on the economic benefits, to buy into energy saving, to understand why this needs to be done and the processes involved. They also need to be told why the issues are important, and how money spent on resources such as comfort cooling could affect their jobs. When people do buy in, the problem goes away to a great extent. This is a fundamental part of achieving a sustainable building, but it is currently neglected, without a doubt. We will look back in 20 years and view this current situation in the same way that we do when we look back on the health and safety arrangements of the past. Senior management has missed a trick on this so far. The issues should be analysed within budgets and finance systems, so that behavioural change can be measured effectively along with physical changes in the building. This could be achieved at nil to low cost, and there can be a very fast payback. This can be directly correlated to elements of the principle of the energy hierarchy as referenced throughout this book. The same principles apply here as within the energy hierarchy itself, and the first and greatest principle is: don't spend on technology to achieve a solution without first identifying and targeting the solutions that can be achieved through behavioural change.

Ultimately satisfying building occupants is the only real test of the success of a building, but only if they are sufficiently part and parcel of the whole process, being motivated and educated to save energy and make this very much part of their work process and thinking.

References

Becker, L.J. (1978) 'Joint effect of feedback and goal setting on performance: A field study of residential energy conservation.' *Journal of Applied Psychology*, 63, 428–433

Guardian (2009) 'Supermarkets get cold feet over fridge doors', 1 October 2009

NFWI (2011) *Close the door on energy waste – Campaign action pack*, The National Federation of Women's Institutes http://www.thewi.org.uk/__documents/public-affairs/past-campaigns-info/nfwi_1211_closethedoor_final.pdf (accessed 21.8.2012)

15 Putting my own house in order

I've always believed that you should practise what you preach and this should apply more so when you're writing a book about sustainable buildings! So if you are in a profession like mine, encouraging people to be efficient and to get their buildings right from a sustainability point of view, you need to be able to answer the question: what have you done to put your own house in order? Knowledge and information about the right things to do are not enough if you cannot demonstrate that you try to apply your own principles to the goals of making a more sustainable society. A lot of what I've written about in this book I've attempted to carry out and implement in my own backyard.

I had a dream

I had always had a dream of building my own house. This was something I had wanted to do from a young age. When I was younger, working on an oil rig and amusing myself in my spare time with plans for the house that I would build one day, I already knew that I would realise my dream in the most practical and as environmentally friendly a way as I could. I didn't know where or when it would happen (this was back in 1983), and it took me another 16 years before I realised this ambition. As I mentioned in the Preface, I finally took the plunge after a family holiday in Canada, when I saw what self-builders could achieve.

As a dress rehearsal to this project, I had already gained some experience by building an extension to a previous house of mine, which gave me an insight into the whole range of procedures and trades that I would have to master and engage in. What was most useful about building the extension was realising what was realistic to attempt. It also gave me useful experience of dealing with the statutory authorities. This made me more confident about dealing with planning and building control.

Delivering Sustainable Buildings: an industry insider's view, First Edition. Mike Malina.
© 2013 Mike Malina. Published 2013 by Blackwell Publishing Ltd.

Jack of all trades

To build my own house, I became, in effect, the client, the architect, the project manager and a whole range of other trades people all in one. I did a lot of the work myself – including a fair bit of labouring! Starting the process meant finding a site first of all. Having grown up in a city, but also having done a lot of travelling and recreation in the countryside, I was always clear that I wanted to live in a very rural environment. Over time, I had noticed that every time we moved house, we went to a more and more rural environment. In terms of a site, we narrowed it down to Suffolk, because it's a unique place.

This area has the combination of a pleasant climate and a great variety of landscapes. The eventual site that I chose, which is on the border with Norfolk, lies between the Fens and Brecks. Most people have heard of the Fens, but fewer know about the Brecks – one of the driest parts of Britain. It is a low-lying heathland eco-habitat, with a unique biodiversity, where at least 12,845 species have been recorded and only found in this area or making up a significant proportion of the UK's population of this type of flora and fauna. In fact, 28% of all the priority biodiversity action plans (addressing threatened species) in the UK occur in this area (Dolman et al. 2010).

It is an ancient landscape, inhabited since the Stone Age, although changed to a fair degree by human intervention. This includes the creation of the largest lowland forest in England; the area is now challenged by modern development, but there are still some fantastic unspoilt areas, including an extensive RSPB wetlands reserve. There is also a site of special scientific interest, created from glacial deposits with some unique geomorphological features left over from the last ice age which can be seen nowhere else so extensively in Britain. As described in the Preface, my wide interests in the natural and manmade environment are certainly catered for in this region.

This area of East Anglia is also home to a number of pioneering best practice low carbon buildings and also has a good concentration of biomass power stations as well as a number of wind farms located in the flat areas of the fens. This includes Ely Power Station which is currently the largest straw burning power station in the world generating over 270 GWh each year.

Further afield in this region, the offshore wind industry has grown enormously to develop and deploy one of the largest offshore wind farms in the world. This will provide major investment and job opportunities in my home region.

Greenfield or brownfield

Originally, most people look for a plot of land on a greenfield site, but this flies in the face of sustainability. Going back to the sustainability hierarchy, which we always need to keep in mind if we are trying to minimise impact, we can see that reuse is always far better than using new materials – and that includes land. From a sustainability point of view, established sites are far better, most of the time, because you are not taking away another natural habitat. From a financial point of view, an established site often has the benefit of established mains services such as water and electricity. The other advantage is that it has already got access and potentially roadways. There is also another pragmatic reason for considering a brownfield site; it is far harder to find greenfield sites.

Careful contemplation and a lot of searching was needed. I was looking for something which either needed complete renovation, which was one option, or which was in such a terrible state such that it wasn't really in a fit state for anything else other than demolition and rebuilding. Another option that occurred to me was reusing some of the materials from any demolished house, where possible. The other thing that, unfortunately, has to be borne in mind is the taxation rules – if you refurbish, you have to pay VAT, currently set at 20%. On a new build, you can claim the VAT back. The HMRC have a scheme called VAT refunds for DIY builders (VAT 431C). This is something that I've always thought was crazy, since surely we should be encouraging people to refurbish, shouldn't we? So the VAT refunds should be a financial incentive for repair and refurbishment. Anyway, we have to work within the law and the rules that exist at the time.

Given this, I decided to opt for the best of both worlds. I found a very dilapidated bungalow and demolished it. I reused a lot of the demolition materials to fit in with my environmental beliefs, but took advantage of the VAT regulations by taking the building back down to its foundations. (If you go down to ground level, it's classed as new build.) Restoring a house would potentially be better ecologically, but the taxation scheme would have worked against me, as the full rate of VAT would have been applied. The whole process was mix of environmental and financial concerns, and as such is a typical sustainability dilemma of the kind discussed in Chapter 1.

The positive feature however, was the possibility, which I explored very early on in consultation with building control, of reusing a substantial amount of the previous foundations, which saved a lot of resources and money. Through pure luck, the owners of the previous bungalow had applied 20 years before to extend it, but the building works had never taken place beyond ground level. What had happened is that they'd cast the foundations, but never taken the project any further. Then they'd moved, and the new owners had let it go to ruin. Severe and extensive damp was present which had bridged the damp proof course where the external soil had piled up against some of the main walls. One of the biggest problems to any building is water and damp, which can cause all kinds of issues. It was thus a severely compromised building. However, I was really lucky that the previous owners had cast those foundations. So my design was on nearly the original footprint, although we were going for a two-story chalet bungalow, Scandinavian style. I wanted to utilise the roof space, as I've never understood why, in the UK, we don't use roof spaces the way they do in many other countries. So that was my intention, which would obviously mean that I needed a new part of the foundation to cope with the wider footprint of the new building; but 75% of what I would need was already in place. I checked with building control, to find out if it would be acceptable for that to happen. We had to dig three inspection pits in sections, so that the building inspectors could actually see the profile and check the depth and quality of the foundations. Several things were in my favour. The geology of the site was well-compacted sandy soil underlain with a solid foundation of chalk. So, after looking at the inspection pits, the building control officer said yes, no problem. I was delighted. Apart from my commitment to reuse, if I'd had to start again, the money that goes into ground works can be unbelievable. The old saying 'it costs a fortune to get out of the ground' is often true.

Architect and planner-talk to the planners

Before I talked to building control, of course, I had had to talk to the planners. The first stage was to find out what they would deem acceptable. I knew this would be helped by the fact that there was already a dwelling on the site, so it was easier than it would have been on a greenfield site, because the precedents for development were there. I had informal chats with the planners as part of the process, going through the processes that I've talked about in Chapter 2. The informal chats were successful, and they said, 'Yes, that's acceptable'. The precedent for the extension, the plot size, the size of the dwelling and the fact that it was a brownfield site were all in my favour. (But I would later find out just how pernickety planners can be.)

The next step was to put my architect's hat on and drew up the plans. This was 1999, when CAD was still evolving and less common. So I spent three days with a drawing board and a T-square, and did the plans by hand as people used to do! Then I filled in the forms and submitted them to the planning authority. The planning was not just for construction; I had to seek official approval to demolish the existing building. When I came to deal with building control, I had to serve a demolition notice to them as well.

I waited for a decision, and while that was going on I drew up a project plan. I always intended to build with a timber frame, adopting a more Scandinavian methodology. I think timber frames are far more sustainable than the traditional brick and block house. This isn't everyone's view, but it's my opinion. The advantage with timber frame is that it's quick to construct, using offsite fabrication. This project was in the earlier days of offsite fabrication, and if I was specifying now I would have ordered the frame with structurally insulated panels (that's with the insulation and some of the services built in). Things have evolved in the past ten years or so, but I had to use what was available at the time.

I looked at what had been done in Sweden and Canada, and how things were constructed in a typical Scandinavian timber framed house. I sought out timber frame companies, gave them the specification and got a price to manufacture and deliver a timber frame. My mission was to construct a shell that I could then work on. Of course, I needed this to be structurally sound and weatherproof as soon as possible.

To follow the programme and keep to timetable, I had planned the process of obtaining structural drawings and calculations for the design and approval of the timber frame, and all I was waiting for now before I signed and placed the order was the official planning permission. I couldn't do anything before this arrived. It would hopefully be a formality, given my informal talks with the planning department but you can never tell for sure. Fortunately, this was not a controversial application. I had thought through consulting the neighbours, and spoken to them before the plans were submitted. They saw the benefit of getting rid of the old bungalow. People often forget this community dimension, but it's very important to get those who will be affected on side. Because the application was not controversial, it was decided under 'officer delegation'. Most decisions are, unless they are potentially controversial or larger scale developments. Nevertheless, some officers have to find something to question! The officer was generally satisfied with the design and approach, which emphasised the environmental credentials. What they can be difficult about is what I would call the finer points of detail and design.

Some planners have a certain view of design and appearance, and unless you satisfy their conditions and make some concession they can block or slow down the permission. One sticking point for me was on the roof layout. I was trying to go for a simple roof with a straight roofline, with no chimney. I had no intention of burning anything on site! But the officer said he didn't like the fact that the roofline didn't have any break in it. He said it was not appealing to the eye. I explained that chimneys were expensive, 'old-fashioned' (my view!) and in my design and function not needed. I wasn't going to burn solid fuel. At one point, he said to break up the roofline by building a false chimney. I was not prepared to do this. The compromise in the end was that he deemed it acceptable to do without the chimney if I altered the roof at the end, so instead of a flush gable end, I put in a barn hip, i.e. a sloped end to the roof to wall elevation. However, there was a cost to this, albeit less than the chimney would have cost me. Planners can cost you money! Yes, the barn hip was potentially more appealing to the eye, but it meant more structural alterations, timberwork and roofing structure, so this cost me money. This, however, was far preferable to the chimney. Other than that, other conditions were details that I had to fit in, including their design guides on external facing and roof materials, which actually fitted in well with my design. I always intended to use clay tiles, as they use less energy in their manufacture and resources than concrete based tiles, and they are therefore more sustainable. On those conditions, permission was given.

Specification, materials and construction

At this point, I could place my order with the timber frame company. At the same time, I applied for building regulations approval, sending in the structural drawings and timber frame structural calculations. I also had to supply information on my SAP calculation, the standard assessment procedure for energy. Under the regulations of the time, which were a lot less developed than they are today, my calculations were fine. Having said that, my specification was ahead of its time, and would still be satisfactory with the regulations in existence at the time of writing. This was a choice that I had made, to put my money where my mouth was, and take out a slightly larger mortgage to fund a long-term higher specification on the timber frame, more insulation, systems and controls. I had worked it out that not only did I agree with it on principle, but it would pay dividends on reducing energy bills in the long run.

Now construction could begin. The next stage was to do the ground works. One of the things I was conscious of was my own use of resources, and I intended to reuse and recycle as much of the previous bungalow as possible. I had the advantage of having 75% of the footprint in place. Also, I had a pile of approximately 170 tons of mixed brick, old roofing timbers and demolition rubble from the existing bungalow. During the demolition, carried out by a local contractor with a JCB, I hired a local villager to help me sort and separate as much of the timbers and useable brick as possible. This I was able to do carefully, because most of the roof timbers were in reasonable condition, and these were later reused for building frames for outbuildings and various other projects, including fencing. We also took all the roof tiles off by hand. This was very hard work, but good for my physical fitness! I systematically put them on pallets, because I had found out that there was a market for second-hand reclaimed tiles and I managed to sell the 70% which we were able to get off the roof in good condition. This broke even for the cost of the labour to reclaim them. It would have been a lot easier to demolish it in one go, but this would have broken my principles of reuse. In the end it was cost neutral but gave me a sense of satisfaction. The remaining 25%, which were not fit for purpose, were added to the heap of broken materials that would eventually be crushed and reused on the over-site for the project. This forms the blinded hardcore, the pre-foundation rubble, which is compacted to consolidate the over-site before casting the floors. So I was saving on materials by reusing materials already on site. This is more common practice now, and there are sites where crushing and screening happens. I also recovered several pallets worth of old bricks for walling and future projects. The remaining 150 tons of demolition rubble, mostly broken bricks, tiles and old plaster, stayed on site for five months, until I found a local contractor who was in the vicinity with a crushing machine, who could take that waste at no cost to be reused on farm tracks and in road material. So I had no landfill whatsoever. That is something I am proud of. Most construction projects result in some landfill. My project didn't. If I couldn't renovate, this was the next best thing, with the materials reused for a purpose, preventing new raw materials being used needlessly.

While I prepared the site, the timber frame structure was being manufactured. However sustainable your build, you still have to use a firm foundation, and the easiest is to use concrete to cast the foundation. Even though I reused as much as possible, I had to import ready-mix concrete to the site. This was not sustainable at all, but there were few other options. This has to be done precisely so that the timber frame fitted. I checked the site dimensions five times to make sure it was right from every angle. The timber frame panels, complete with waterproof breather membrane, arrived on the back of a flat-bed lorry, and within five days, which was an amazing thing to watch, the timber frame was erected by an expert team of contractors. The roof structure was also done as part of the process. It was for me to put on the roof insulation and, over that, the waterproof roof breather membrane, and to get it as weather-tight as possible. Being a typical British summer, it rained and rained, which was also ironic as this is one of the driest areas of the UK and is often prone to drought (more on water conservation later). So I spent lot of time sweeping out water in between working on the roof. The roof structure that I had designed is what is known as a warm roof – so you insulate the outside of the timber structure to utilise all the roof space as accommodation. This I did with rigid board insulation. This is very expensive, but pays dividends in long term as you get very good U-values for the thermal properties of your construction. By chance, I happened to be reading a self-build magazine at the right moment, and saw an advert for a company that sold 'seconds' of insulation board. When the boards are made, if they have cosmetic faults – say, they are undulating or slightly tapered – the manufacturer rejects them. They don't sell these 'seconds' boards in normal builders merchants. This company took these boards and sold them a lot more cheaply than normal. So I ordered the entire batch to apply to the walls, floors and roof and saved a lot of money. It's madness not to use them – there is nothing wrong with the board, and you won't see it again as it's inside the roof and walls. There is an obsession in modern society with perfection, from goods in supermarkets to products on site. Everything needs to be cosmetically perfect. We need to be less intolerant of the quirks of nature and of some general cosmetic faults in this type of production. That was another example of how I utilised material that might otherwise have gone to landfill or waste. Also, I always had it in mind that being resource conscious saves money as well as the environment. This is a good case in point.

Skills and knowing your limitations

The next step, once the insulation was in place, was to get the roof tiles on. It's at this point that I realised my limitations. I will have a go at anything, but there comes a point where you realise that some tasks are very skilled and should be done by an expert. I have found out during construction, that you really need a professional roofer. So I hired one and helped as a labourer. I learned an awful lot and realised that setting out a roof is a skilled task.

At this point, I was labourer, learner and ultimately project manager. Also, I was the main contractor, who employed subcontractors. As well as employing a roofer, I hired some labour to work with me on a number of things, and also a brickie. The only other thing I needed to employ someone else for was plastering. Being the person I am, I had to have a go, and this is how I found out how skilled you have to be to plaster a ceiling. What I found I could do was put the plasterboard up. This is straightforward but hard work. Plastering I could actually do to a point, but not at the scale and rate of a professional plasterer. This gave me such an insight into the construction process – when it's your own project and your own money, you become more conscious of everything.

Mechanical and electrical

My core skills were the mechanical and electrical elements, so I did that entire part of the project – all electrical, heating and plumbing input throughout the build. This was pre-certification, pre-Part P of the building regulations and the other building regulation approvals, but I still had the electrical installation checked by a professional electrical colleague and I'm glad to say it got a clean bill of health. My whole approach to the mechanical and electrical elements was very much on the same philosophy that I've written about in this entire book, which is basically to build in the best future-proofing and the best possible standards of installation I could, fully designed with maintenance in mind. The advantage of doing it myself was that I could build-in virtually total accessibility to all the pipe runs. To show how technology has moved on – I also ran CAT 5 computer network cable throughout the house, thinking I was future-proofing for my computer needs. Of course, with the march of technology, wireless and the ability to use mains electricity cables for electricity and data exchange are now readily available. Trying not to be defeated, because I didn't want the cable to become obsolete, I realised that the cabling could still prove useful in an IT infrastructure and potentially for any future controls network that I might choose to upgrade within the house.

The other thing about building from scratch was that I was able to future proof electrically. I got the largest consumer unit I could get to try and break down as many circuits as possible for the future. All my lighting was to be as energy efficient as possible, and at the time, in 1999, compact fluorescent lamps (CFLs) were still relatively new and have since undergone a complete revolution in their design and availability. They have also got a lot cheaper, but I bought early-generation CFLs and have several that are still working fine 12 years later, which proves that the technology lived up to its initial expectations and has delivered some fantastic cost savings over traditional, conventional lighting. I once tried to work out the savings over a decade, and my rough calculation was that it had saved me well over a thousand pounds in electricity costs and consumables. That's an example of early adoption which shows that sustainability pays. I'm now experimenting with LED lighting which, I believe, will transform the way we use lighting over the next decade and will have a major role to play in conserving resources and in energy efficiency.

The heating system I installed is under-floor, which is fed by controlled manifolds. Each zone is individually controlled. At the time, I fitted basic thermostats because the cost of programmable digital thermostats was so high. But I always knew that one day the technology would become more commonplace and the costs would come down, so about seven years later I changed them all. This enabled each room to have not only its own temperature control but a programmable time control also. This is what I've advocated for commercial buildings for a long time, and they are available for domestic buildings much more cost effectively now. This is a good example of allowing things to be upgraded when the technology becomes available. The development of an automated and highly controlled building environment should always be a priority. One of the things I spent a long time on was pre-commissioning, commissioning and balancing the heating system, making sure that each zone was perfectly proportioned and balanced in terms of its hydronic flow. This was easy to achieve, because all the manifolds have a flow measurement setting. In terms of pre-commissioning, I thoroughly flushed the system made sure that there's an appropriate corrosion inhibitor, which I maintain myself.

Most plumbers still use copper throughout their pipework, although there has been a greater use of plastic barrier pipework and plumbing in the past ten years. I made a decision back in 2000 to use barrier pipework and plastic plumbing throughout, weighing up the pros and cons from an environmental point of view. I concluded that plastic was better than copper, albeit the lesser of two evils. I've always thought that, although it's taken time for the plumbing industry to change, barrier and plastic has a lot of advantages in its corrosion resistance and ease of fitting. One thing I hate with copper is when you don't get the joints right and have to drain it before you can re-solder it. With plastic plumbing, you don't have to drain it down, you can do it as a wet system. The argument that a lot of plumbers make against it is that it's not time proved, and that they like copper because it's solid. I had this discussion 15 years ago, and I have to say that (touch wood!) so far I have had no problems whatsoever with over a kilometre of pipework. There is another issue of course – copper is very expensive and contains a lot of embodied energy, more so than ever now.

Sustainability dilemma

As covered in Chapter 1, I had my own sustainability dilemma: what is realistic, what is technically possible and what is the cost? After all, mortgage limitations are a big consideration to the reality of what can and cannot be done.

The source, type and specification of materials is important in delivering as sustainable a building as is practicable. It is also subject to much opinion and debate. I tried to design and specify the best possible, using what I considered the best materials available to give me the least environmental impact and lowest carbon footprint. Part of this was trying to ensure that the materials came from as sustainable a source as I could find. It is difficult to realise total carbon neutrality. I don't believe that, at present, we can source and realistically achieve true zero carbon. It's more a case of trying to be as low carbon as possible. Carbon is embodied in the raw materials and resources needed for the manufacturing process. The material has also been transported, which will almost certainly be by a carbon fuelled vehicle. One principle I stuck to was to ensure that all the timber was obtained from a sustainably managed source. The Forest Stewardship Council (FSC) 'tick-tree' logo was a useful mechanism to ensure this. This applied to the timber frame, internal studwork and the window frames.

One of the key issues which merits its own chapter in this book is water – Chapter 7. This is a key issue in sustainability and indeed a recurring theme. As mentioned previously I live in one of the driest parts of the UK. This is also an extensively agricultural region and community. Water is therefore a vital issue. One thing I can certainly say is that no rainwater is lost to the mains sewer. All rainwater is captured to large water butts with an overflow to soakaways to return the water naturally to the water table. I have also devised a means of diverting a significant amount of rainwater to a large pond that I created, which has become a haven for an incredible diversity of wildlife. I have investigated the possibility of utilising rainwater for flushing toilets or even clothes washing, but the technical issues and costs associated with this meant that it fell off my energy hierarchy agenda. It is again about making choices, and I have put considerable effort into testing flow restrictors and other devices to some protests from my family. See Chapter 14 on behavioural change!

The future: improvement and continuous commissioning

So what am I doing to improve the house? There are a number of items that I would have liked to have done and included in the original design and specification. If I'd been undertaking this project now, over a decade from the construction, I would have had access to a range of technologies and materials that were not readily available then or that were prohibitively costly. The problem now is determining what is economic to retrofit, and this highlights the importance of the energy hierarchy. I followed this philosophy in ensuring that I designed the building to minimise the need for energy and then to use energy as efficiently as possible. I could have increased the amount of insulation even more, but there is a point when a line has to be drawn; diminishing financial returns on investment or payback weighed up against the embodied energy in the insulation and the effectiveness of the insulation itself. The advent of the renewable heat incentive (RHI) and feed-in tariffs (FIT), coupled with the green deal could provide incentives to investigate other possibilities in reaching the final stage of the energy hierarchy. The idea of being able to generate electricity and perhaps rely less on the national grid, seems appealing. The costs on the other hand are always a consideration, as are the issues of embodied energy and the relative inefficiencies of the current technology. Solar PV is still at best only around 15% efficient on conversion of sunlight into electricity. I'm sure this technology will progress and this is something I will investigate, but at the time of writing, the government seems in turmoil and conflict with the PV industry as uncertainty reigns with the FIT.

Managing the electricity grid is something that we can all help with. I am hoping that the opportunity to use dynamic demand technology will become available and hopefully all fridges, freezers and non-critical equipment will be able to use this, helping to reduce energy at source. I have fitted a domestically available voltage optimisation unit and that has delivered an 8% reduction on my electricity use.

I am particularly interested in the possible use of fuel cells in the future and the greater use of thermal stores, which may become more efficient with the development of phase change materials. Technology changes and develops and we are in for some challenging and exciting times ahead. My own step towards a lower carbon future and my attempts at moving towards the goal of sustainability including the building of my own house has been a fantastic experience, not only from a personal point of view but also professionally. I hope that my two children, who have grown up now, have benefitted from being immersed in this environment. They are certainly a lot more aware of some of the key issues of sustainability and its developing role in the future. They have certainly carried forward the childhood experiences of living through a self-build project and remind me of this fact on occasions!

The one thing I can guarantee in this industry is that you never stop learning. Also, technology moves on, but the basic principles remain the same. This is all good for skills development and building towards this sustainable future. What I'm looking forward to now is being able to deploy even more energy efficient technologies and ways of reducing my impact on the environment – and my pocket! Anything that achieves this is a win-win situation.

References

Dolman, P.M., Panter, C.J., Mossman, H.L. (2010) *Securing Biodiversity in Breckland: Guidance for Conservation and Research. First Report of the Breckland Biodiversity Audit.* University of East Anglia, Norwich.

Further information

For my part in promoting East Anglian tourism – please visit:

www.brecks.org
http://www.brecks.org/visitor-guide-brecks.aspx
www.visiteastofengland.com

16 Sharing our technology and expertise with the developed and developing world

In the developed world, we take complex buildings with fully functioning building services for granted. The level of mechanical services we are so used to has its roots in 19th century innovation, but with the advent of electrical development and infrastructure, the technology became more widespread and became established and embedded in our culture and standard of living in the 20th century. We no longer think twice about the fact that we can just flick a switch to get electric power or light, or just turn on a tap for water, including hot water if we want. However, this is a level of comfort and convenience which is utterly unknown to vast numbers of people on the planet. As I finish writing this book, we are beginning the 2012 United Nations International Year of Sustainable Energy for All (UN 2012), yet most people I speak to know nothing about it. By the time this book is published, the year will probably be all but finished, yet this issue doesn't finish here and is in my view one of the most important issues that faces the future of humanity.

The big picture

I had the privilege of travelling in Africa in my early twenties, and I saw some real poverty and basic living. I also saw some real innovation and clever application of generations of passed down knowledge in how to live and survive in very hostile environments. Africa is full of sights that make you think – on an intellectual level, we know that some people are carrying water for miles in order to survive, but being confronted with the reality of children walking miles every day carrying water is different. I also later travelled in India and South East Asia, and again saw some very basic modes of living. It's humbling to witness the ingenuity with which people utilise their limited resources to good effect. I've seen people collecting and stacking cowpats to dry for fuel. Collecting brushwood is also common, and it is time-consuming work to fulfil a basic need. Almost half of humanity relies on biomass for cooking and heating (UN 2012). This way of life will be unknown to most, if not all, readers of this book. The nearest we

Delivering Sustainable Buildings: an industry insider's view, First Edition. Mike Malina.
© 2013 Mike Malina. Published 2013 by Blackwell Publishing Ltd.

come to an idea of how the majority of people live is on the rare occasions we lose water or electrical supply. When the lights go out or the tap doesn't run with water, it gives us a small insight, for a fleeting moment, of what life would be like without these resources. The rural area where I live underwent a three-day power cut as a result of winter gales that brought down many overhead power cables. I utilised the car engine with a small electrical inverter as a makeshift generator to create mains electricity (within the confines of safe protocols), to power basic lights and the heating circulation pump, but operating that was limited by the availability of petrol. That was inconvenient and of short duration, but I still had more access to power than the two billion people in this world who don't even have any access to electricity. Those in the developing world that do have access often have to put up with a disrupted supply on a regular basis. Even though so many people have to live with this situation, I saw in India some amazing creativity in the back streets of some of the towns and cities. People were showing some incredible engineering skills in manufacturing spare parts and fixing nearly everything, even for example, filing gear wheels by hand. This emphasised the value of resources to people who have much less than many in the 'western world'. We have forgotten the importance of resources and have evolved into a throw-away society. This has to change if we are to achieve anywhere near the goal of a functional, resource conscious sustainable way of living.

Losing water supply in the UK is a very rare occurrence, and usually a temporary problem caused by maintenance works. There have been a few instances where a water main has broken, but in the developed world, we

Figure 16.1 Carrying water for miles (credit: Practical Action)

only suffer a little bit of inconvenience. We can still get our bottled water or perhaps walk a few yards to a water tanker. By contrast, people in the developing world walk maybe 5 miles every day to get water which may not even be safe to drink (see Figure 16.1). Imagine once or twice a day walking miles to get water. If our infrastructure collapsed, how would we cope? We can survive without electricity, but without water it's a different story. Try going without water even for a day – you can theoretically last longer, but it's not advised. The UN has estimated that 1 billion people have no access to safe water, and also that by 2050 there will be 30 million more people who may be hungry because of climate change; 1.5 billion people have inadequate shelter; 1.6 million people die each year from the effects of burning fuels in their houses, from the indoor smoke. For us in the developed world, facing these living conditions would be a huge culture and practical shock – it could lead to the total breakdown of society as we know it.

Sharing our expertise

We've all seen on the television and in the newspapers, images of famines in Africa, of catastrophic floods and earthquakes around the world, and even manmade disasters from the conflicts of war. This spurs many people to donate to the numerous agencies working to relieve these situations. Charitable giving is good, but it can be superficial. It too rarely tackles the root causes of the problems, so they can reoccur generation after generation. One thing that has always interested me is how we can go beyond digging in our pockets for short-term relief and move to the creation of longer-lasting solutions.

All UK governments in recent years have had an aid budget. Aid is one of the marks of a civilised nation and society. Historically, the UK government set up the Department of Technical Cooperation in 1964, to deal with technical side of an aid programme and to coordinate government action that was previously spread between different government departments. Also in 1964, the Ministry for Overseas Development came into being. Today, we have the Department for International Development. All of these government bodies have indirectly involved us all in funding aid through our taxes. In 2010/11, the UK's gross public expenditure on development was £7.7 billion pounds, which is just over 1% of UK spending (DFID 2011).

There are a considerable number of non-governmental organisations (NGOs) as well. One example would be Médecins Sans Frontières (Doctors Without Borders), who make a real difference to the developing world by providing much needed and vital medical treatment and care. Their programme deals with both short-term and longer-term measures. As well as basic provision of care, they have extensive education programmes, which build capacity for the longer term. Tackling the issues of development needs a multi-pronged approach, so education is vital, as is the need for access to information about technology and contraception. Otherwise, large families perpetuate the spiral of poverty. Part of any good aid programme should also be about providing energy and decent shelter. A lot of this happens through developing-world trusts and aid charities, and from the overseas development budget. But there are also many sector organisations such as Engineers Without Borders, EngineerAid, Architects Sans Frontières and Architects Without Borders.

Like many in our profession, I am drawn to the idea of doing something practical. We're in the game of making things happen from a practical and engineering point of view, so we're well placed to find solutions to developmental problems. The collective skills and knowledge of our industry is vast, so we have the potential to achieve much. Apart from benefitting others, professionally it would do many of us a lot of good to look at technology transfer issues that take us back to basics in terms of our knowledge base. We are now so technologically developed in the West that we're in danger of losing touch with basic engineering concepts. BEMS engineers might be experts at software systems and building control, but that skill-set would be totally useless to someone in the developing world who has a makeshift shelter, where they're interested in looking at the basics of heat, light and water. Nevertheless, that same engineer, as a skilled practitioner, would have a much better idea of the basics, and, with thought and application, could potentially train and transfer knowledge to the developing world, whether practically or in a training role. I have long taken an interest in and been involved with the organisation Practical Action. It was previously known as Intermediate Technology, and as this name suggests, the organisation is focused on knowledge transfer of appropriate-level technologies that can improve people's lives. Figure 16.2 shows a practical message. It's an excellent message that conveys practical knowledge and skills transfer – making a functional plough from scrap metal to make a blade perfectly suited to local conditions. Figure 16.3 shows the deployment of an effective hand pump to draw water from a well.

My interest in their work goes back to the mid 1980s, when I used to devour the basic technology books in their London bookshop. This bookshop was full of practical books on an incredible range of practical technologies, such as solar water heaters made using basic materials, hand pumps, irrigation and farming techniques and basic wind technologies. All these 'intermediate technologies' can make such an impact on these communities in the developing world and also build local skills and education to enable these communities to develop a sustainable future for themselves. More information is available at: www.practicalaction.org

This kind of intervention has to be understood within its historical context. Firstly, there was the impact of colonialism and its imposition of western ways of doing things. This happened all over the world, with groups from the Australian Aborigines to the Zulus forced to give up their traditional knowledge and adapt western 'civilised' behaviours. Many skills for survival and sustainable living were lost. Secondly, there is the issue that the newly independent nations that rose out of the colonial era were quickly saddled with debt related to weapons supply and internal struggles from artificially created countries with boundaries that crossed historic tribal lines. This

Figure 16.2 Practical action – a great message

Figure 16.3 An effective hand pump to utilise water from a well (credit: Practical Action)

diverted resources away from the development of essential infrastructure. While this was going on, more western practices were adopted, such as moving to cities, with population being concentrated in relatively small urban areas. Over succeeding generations, more skills were lost relating to agriculture, basic sustainable building and survival. This 'progress' also led to more fuel consumption, so small farmers were growing cash crops to fund debt. This makes any future development that helps people to rediscover their skills more vital than ever.

This is not just a developing world problem. Many people in the developed world would soon perish if the convenience of the 'modern world infrastructure', underwent any prolonged disruption. We have seen this to a lesser degree, even on a short-term basis with any natural disasters or loss of electrical, gas or water supply, especially in urban areas. Basic skills have been lost over succeeding generations and this has never been so evident when looking around us with the complete convenience of the modern way of living. So what would people do, and how many have a 'plan B' or backup option?

What did the Romans do for us?

If we go back to the peak of the Roman empire, we can see that, compared to their neighbours, they were far ahead of their time, culturally and technologically. Their engineering prowess and skills were remarkable, and they had incredibly skilled artisans, craftsmen and engineers. The 2000 year old Coliseum in Rome displays a level of building engineering which is astounding given the technology and tools they had available to them. From a building services point of view, the Romans developed and learned how they could also control their internal environment. They had running water through latrines, for instance, as well as the famous Roman baths. They had under-floor heating. Figure 16.4 shows the extent of the under-floor heating at Roman Wroxeter, Shropshire. They were very culturally aware of cleanliness and hygiene and used natural ventilation to good effect. There is archaeological evidence for this all over the extent of their ancient Empire – the remains of the infrastructure have been found from Eastern Turkey and North Africa all the way to Northumberland and the communities along Hadrian's Wall. All their achievements were developed without electricity and other modern technology. The Romans knew how to use and harness nature. Had their empire lasted beyond the 4th century, I wonder if they would have embarked on an industrial revolution and made the same advances and perhaps the same mistakes as us. Evidence of technological development is also apparent across the ancient world from the Mayans in Central America, to the Ancient Egyptians and Chinese to name but a few. Many ancient civilisations also had a good understanding of the natural and built environment. We have a lot to learn from our ancestors; they didn't require the complexity of the mechanical and electrical technologies that we take for granted today. They used gravity very effectively

Figure 16.4 My visit to Roman Wroxeter – What have the Romans done for us?

and developed water-powered mills. Many of these skills were lost in the Dark Ages of history and over time, and we are only just rediscovering many of them. The existence of some of this knowledge, and how the techniques were used and deployed are being proved through experimental archaeology. Figure 16.5 shows a complete Roman villa constructed using, as accurately possible, the original roman techniques. Yet we still don't fully understand some of the methods and techniques that were used to achieve all their various practices. Nowadays, many of those techniques could be used very effectively to help the developing world, as we redeploy their skills and knowledge to enable communities to put themselves in a much better position to guard against the effects of climate change and possible future economic changes. With these practices and skills we would also improve our own knowledge and it would make us value and respect our own natural resources a lot more.

Business and exports

So far in this chapter, we've been talking primarily about technology transfer and aid, but there is also a huge opportunity for companies to work with emerging economies. In particular, businesses should note the development of BRICS. This is an international group comprised of Brazil, Russia, India, China and South Africa, who are all set to become major economies this century. These states are also bridging the gap between the West and the smaller countries that are less developed, and they will be prime countries for growth

Figure 16.5 Experimental archaeology – reconstructed Roman villa

across the globe. The concern is how sustainable can this be? Population is increasing markedly in these countries – except for Russia, where the population is falling, largely due to the poor health of the population. Overall, these countries will be massive growth markets. The UK currently imports a lot from China; in fact, in 2009, China became the world's largest goods exporter. The UK currently exports more to the Republic of Ireland than to the total BRICS combined but this is surely set to change, as Europe is very likely to do more trade with BRICS. As the BRICS economies grow, they are consuming a vast quantity of natural resources. China is consuming the greatest amount and, because of this, the Chinese are going to South America and Africa to establish bridgeheads where they can exploit natural resources. This in turn is having and will continue to have a massive impact on these continents' economies and population as they exploit their resources to sell to China.

There is also a wealth of talent within these countries. China, India and Brazil are producing a large number of graduates to help their developing economies. This is already happening in India, for example, fuelling technology change and economic growth. We see now, for example, the number of Indian graduates who are going all over the world as there are not enough jobs to go round in their own country at present to utilise their range of skills. They produce a lot of computer and engineering graduates. This has been a boon for British universities, because so many of them trained in the UK – another form of technology and knowledge transfer. Indeed, we should see this process of up-skilling and growth as a massive opportunity, because we will be in a position to export a large amount of expertise and technology to help these developing economies maximise their efficiency and help them towards a path

One of the things I keep seeing is trade missions. The UK government set up the UK Trade and Investment (UKTI), which brings together the work of the Department for Business, Innovation and Skills and the Foreign Office. Part of their function is setting up trade missions for inward investment and export, and there particularly seem to be a lot of initiatives aimed at BRICS. I have received a number of invitations from them. For example, there was a visit in March 2012 made up of a delegation from the Brazilian Chamber of Construction. This aimed to showcase British expertise in green and sustainable construction. My view is that this is the kind of opportunity we should be embracing in the future.

Figure 16.6 Left to right: Eleni Vonissakou, Commercial Officer, British Embassy; Anastasia Marinopoulou, Business Development Manager, BSC; the British Ambassador, Dr David Landsman and Mike Malina, ESA

UK embassies have trade missions or specialists whose job it is to secure contacts for British companies. Many are working on sustainable construction and green technologies. My own experience of this was a trade exhibition in Greece, where the UK government funded the hotels and helped with exhibition costs, providing a UK stand from which British companies could exhibit and base themselves to make networking and presentations possible. (Figure 16.6 shows the author with colleagues and the British Ambassador to Greece.) This is an example of government supporting British business to exploit knowledge transfer for commercial opportunity.

of sustainable growth. We need to influence the way the new economic powers grow to ensure that it is sustainable. I'd like to see a lot more academic and technological collaboration, using the best technology and scientists to find more effective ways of utilising scarce resources and of using low carbon technologies. This is starting to happen, but needs to happen more.

A classic example of opportunities for this kind of technology transfer would be with HVAC and building management systems. Significant numbers of buildings in BRICS are being built or upgraded – these are enormous emerging markets for UK companies to export the hardware and software to help create sustainable buildings. Ultimately, these markets are perhaps a hundred times greater than our own. We don't have to look internationally from a purely altruistic perspective; there is an enormous opportunity to make money through the creation of sustainable growth. If we can make wealth as we assist, everybody wins. There are several reasons as to why we should be sharing technology and expertise. Firstly, there is the business opportunity to sell to a very large developing market. Secondly, it is in our collective interest as a species to assist each other to manage our environment and grow as sustainably as possible, because BRICS and other nations are going to have more of an impact on our climate than the entire impact to date, due to the fact that BRICS and other developing nations encompass more than half the world's population. We cannot say that they cannot have the standard of living that we have, but we need to help them to deliver change in a sustainable way. Thirdly, it's our duty to try to share the best technology and techniques to help them to achieve sustainable growth, for the sake of both their development and the human race's future survival. This is clearly in all our interests.

References

DFID 2011 Department for International Development http://www.dfid.gov.uk/About-us/How-we-measure-progress/Aid-Statistics/Statistics-on-International-Development-2011/SID-2011-Section-3-How-much-is-UK-Expenditure-on-International-Development/ (accessed 22.8.2012)
UN 2011 international year sustainable energy for all www.sustainable energyrall.org
UN 2012 United Nations Foundation http://www.unfoundation.org/what-we-do/issues/energy-and-climate/clean-energy-development.html (accessed 22.8.2012)

Further information

Practical Action: www.practicalaction.org
Engineers Without Borders: www.ewb-uk.org
EngineerAid: www.engineeraid.com
Architects Sans Frontières: www.asf-uk.org
Architects Without Borders: www.architectswithoutborders.com

Conclusion – some big challenges ahead

It is difficult to overstate the enormity of the challenge posed by climate change, resource depletion and energy demands, with the need for all of us to embrace a more sustainable lifestyle. As previously stated, the idea that we need to 'save the planet' is mistaken: the planet will survive as it has always done. What we need to do is save ourselves. The question is, can we do it in time?

That's a very hard question to answer. As a species, we are clearly capable of great things. The first recorded powered flight took place in 1903 when the Wright brothers made a 50 m flight in a simple biplane made from wood and canvas. Sixty-six years later two men walked on the moon. On the other hand, we have also witnessed catastrophic failures. Easter Island, perhaps most notably, serves as a grim warning from history of how short-sighted and destructive humanity can be and this is echoed in modern times with examples of our destructive impact on the natural world and environment.

The planet is complex, and we're a small part of its overall journey. While wasting energy, it seems, too many of us have lost touch with the skills of our ancestors, and basic knowledge has been obscured by the demands and complexities of modern life. People were a lot more self-reliant and adaptable in the past. It used to be vital to have a good rapport with nature to survive. In the modern era, we're not so close to our natural environment as we were. Skills have been lost. So the priority has got to be retraining and up-skilling and using building services technology to keep our built environment healthy and efficient in a way that minimises the impact on the natural world and its resources. It makes sense in every aspect, whether purely from an economic or business perspective, or on a wider view of how we want to share our continuing existence with all the other people and species on this planet. It will require a big culture change, and culture does change with time. We have seen issues treated as add-on in the past – such as health and safety and disabled rights – become mainstream over time, and I cannot see how sustainability in the built environment won't become one of, if not the biggest, issue of all.

So, if we are capable of putting a man on the moon, can we rise to the challenge of making the transition to a low carbon economy and society? Is it possible for us to make equal progress and show equal commitment as we did to the Apollo projects? Sustainability is certainly a much bigger task than Apollo and is very different in that it involves and affects everyone on planet earth. In fact, the challenge is monumental but, as an optimist, I would say it is still achievable. Nevertheless, it will require a radical change in society and the way we conduct our day-to-day lives.

From the point of view of the building engineering services industry, we will have to play our part by accepting that all buildings, old and new, will be crucial to a successful transition. Our sector will be at the forefront of the challenge. We have the potential to provide a successful blueprint for change for others to follow. On the other hand, if we fail to rise to the situation, the efforts of many others in society will count for very little. The built environment, and its management and adaptation, are crucial to the whole issue of sustainability and human society.

Another concern is that the divide between the more developed and the developing worlds will be cemented and reinforced by the need for change to cope with the impacts of climate change and energy and natural resource demands. Can we, in the developed world compromise our lifestyles and use less so that the developing world can have more? To succeed in this challenge, we need to adapt and export our best technologies to make the developing world as efficient as possible.

Therefore, we need to change the industry and the way we work and interact within our own sector and beyond. We need a new and integrated approach. Making connections between all the issues discussed in this book and building a bridge between wider sustainability issues and sustainable engineering will give us the momentum to do that. This has got to be the way forward.

However, the changes we need, to make our existence on this planet sustainable, must happen against a background in which things are already changing rapidly in other ways. One of the trends which is already evident is the move towards more generic engineers and engineering practices. Clearly, in order to succeed, we will all need a much wider understanding of mechanical and electrical services and a generally much wider brief. Far more more consulting engineers, like me, are moving towards running their own business, so we also need a better understanding of commercial issues, and also the kind of self-reliance which I first developed in my days on the oil rigs. Therefore, traditional engineers and contractors alike are currently adapting rapidly to a changing market, which both reflects and extends beyond the challenges created by ecological issues.

These changes are here to stay. In particular, sustainability is not just a fashion. It demands a change in everything we do. Even although I'm an optimist, I have to admit that achieving the required change in time to save ourselves is not going to be easy. The composition of this book reflects this: I've tried to cover a wide variety of issues, because the challenge before us requires that we gain a holistic understanding of both the technologies and the finance behind it, which I have covered in some detail. Even so, this is not enough on its own, because we also need to look at some of the traditional dimensions of our jobs that lately haven't been at the forefront of our professional priorities: commissioning would be a good example of this. Ultimately, the whole lifecycle of a building needs to be considered, otherwise we will only create substandard buildings which don't work and are energy inefficient.

These issues won't just get solved by new technology alone. We have to engage and work with technology and people to ensure that building services are understood and that they are commissioned properly and work. We also, importantly, need to avoid some of the green bling that we've seen in early adopters, getting carried away from the core issues, neglecting to tackle the more important items in the energy hierarchy. Most importantly, we need this holistic view to create a culture and way of bringing together and integrating all of our practices and systems to improve our existing building stock. The main thing to bear in mind here is that 80% of the buildings that we will have in 2050 have already been built. We're going to have to learn to retrofit and upgrade these buildings significantly. We will have to adopt an approach of climate change adaption to cope with this programme of work. This is the only way we can create a more sustainable and efficient society.

Whatever the pace of technological change, how much intervention will be needed to make this happen? There is an ever increasing amount of regulation and legislation and standards, but if our governments in the future don't start to exert significant enforcement of these laws and regulations, we will not reach our goal as a society. Without compulsion, change will not happen because, in the free market, money talks and people won't do things that cost more unless there's a 'big stick'. However, taking a longer-term view, the growing green economy has the potential to benefit all of us in so many ways. More government intervention is essential to get us moving in the right direction, with a longer-term and more risk-adverse attitude needed. For those businesses that respond positively to the challenge, the economic benefits will be greater profitability and job creation.

Ultimately, everything in this book is about relationships and interconnections. It's about key items such as behaviour change incentives, financial or otherwise, and about how law and incentives are regulated and enforced. It's all about best intentions, about how politicians and people would like to do things. We've got a mixture of policy schemes, e.g. central and local authority schemes, and enough information to fill Noah's ark with books on sustainability, but how do these all combine to create a reality of making things happen? At the same time as I think there is a big disconnect between policy makers and practitioners, I also think that there's a gap between the consumer rhetoric (what people are saying) and the actual performance of the country as a whole on sustainability issues. People get fed a lot of questionable information and green bling, causing them to put their trust in the wrong places – for example, companies promising financial returns on energy saving or renewable products that just don't stack up, either financially or from a sustainable performance point of view. I believe that the only way to get over this is to have a vigorous set of proofs that these returns and investments actually work. These need to be independently verified and trusted, with a common set of recognised and accepted methods to prove that things do as they are claimed – in short, they need to do 'what it says on the tin'.

The only way that is going to happen is through massive government intervention and government and industry inspired standards. We need an enhanced trading standards type organisation, with an accredited badge for things that work; if things don't work then those items should be exposed and any persons responsible for making misleading or false claims should be prosecuted where appropriate. We need to make this highly complex issue, with so many variables and choices, as transparent and as simply explained as humanly possible. Ultimately, we need to be open and truly accountable.

From an industry point of view, I believe the way forward is to adopt and embrace sustainability standards with comprehensive checklists or overviews/flowcharts that make it simple for everyone to follow. Ultimately there is a balance to be reached between freedom of choice and the collective need to do something substantial in preserving our standards of living and protecting the natural environment and the resources we all depend on. How much the state has to intervene to guide this process may be a contentious issue but it needs to be tackled.

I have tried to set out some of issues and choices in this book and I hope that it goes some way towards achieving the goal of stimulating debate and making the links between building services professional, environmentalists and other interested parties. Perhaps it also raises a lot of other questions as well. None of us have all the answers; if we did, I wouldn't be writing this book! As an industry, those of us directly concerned with designing, constructing, commissioning and maintaining the built environment, need to find better ways of communicating and making the links. Education and raising awareness will be vital in taking these issues forward. After all it clearly is in all our interests both professionally and personally to try and make this happen. I hope this book is a small step forward in helping achieve this process.

Recommended reading

There are two books I recommend to any building services or facilities management professional who is interested enough in the wider environment to have gone to the trouble of reading this book. The first is the *Gaia atlas of planet management*, edited by Norman Myers and published by Gaia Books, and the second is *Global issues, an introduction*, written by John L Seitz and published by Wiley-Blackwell, which tackles the links between such issues as geography, food, wealth and poverty.

Index

thermal elements, 42, 43
thermal imaging, 182–97
thermal insulation, 29
thermal solar, 16, 63, 206
thermal store, 231
thermostat, 4, 73, 99, 133, 147, 164, 229
tidal power, 75
timber frame, 224–7, 230
Town and Country Planning Association (TCPA), 33, 34
transformer, 186
transport, 9, 21, 31, 51, 69, 71, 73–5, 84, 85, 97, 99, 125, 230
travel, 53, 73, 120, 217, 222, 233
Treasury, 98, 209

UK Accreditation Service (UKAS), 64
UN Framework Convention on Climate Change (UNFCC), 37
under-floor heating, 133, 238
UNESCO, 102, 113
Unfair Trading Regulations, 17, 24
United Kingdom, 41
United Nations, 233, 242
Uranium, 75
urban, 7, 28, 75, 109, 131, 136, 137, 140, 141, 150, 238
U-value, 227

vehicle, 6, 17, 43, 94, 230
ventilation, 1, 7, 8, 39–42, 63, 128, 131, 135–45, 147, 153–5, 158, 164, 178, 213, 238

cross flow ventilation, 138, 139
volt, 10, 18, 19, 31, 32, 59, 60, 109, 123, 204, 206, 231
voltage, 97
voltage optimisation, 231
volume, 24, 91, 113, 126

Wales, 29, 39, 41
walking, 197, 233, 235
washing machine, 91
waste, 6, 9, 10, 16, 19–21, 23, 28, 34, 39, 40, 43, 49, 71, 74–7, 84, 88, 95–9, 105, 109, 116, 124, 125, 129, 135, 136, 149, 151, 154, 158, 180, 197, 198, 206, 215, 216, 219, 226, 227
waste hierarchy, 124, 197, 198
water meter, 99–101
watt, 59, 60, 76, 77, 85, 128
Watt, James, 77
weapons, 236
wind turbine, xi, xviii, 18, 21, 124
wind farm, 25, 75, 222
window, 5, 6, 8, 43, 93, 96, 120, 133, 136, 137, 139, 141, 153, 164, 165, 230
winter, 6, 7, 113, 137, 154, 234
wood, 31, 135, 229, 233, 243
world, 1, 3, 36–8, 76, 82, 84, 97, 99, 111, 113, 175, 197, 222, 223, 233–44
Wyatt, Terry, vii, viii

WILEY-BLACKWELL

Other Books Available from Wiley-Blackwell

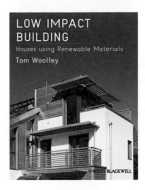

*Low Impact Building:
Houses Using
Renewable Materials*
Woolley
9781444336603

*Ecosystem Services
Come to Town:
Greening Cities by
Working with Nature*
Grant
9781405195065

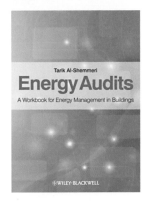

*Energy Audits:
A Workbook for Energy
Management in Buildings*
Al-Shemmeri
9780470656082

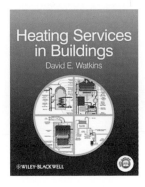

*Heating Services
in Buildings*
Watkins
9780470656037

Sustainable Refurbishment
Shah
9781405195089

Green Guide to Specification
4th Edition
Anderson & Shiers
9780632059614

www.wiley.com/go/construction